Technology Transfer
A Realistic Approach

Technology

Transfer
A Realistic Approach

Silvère Seurat

Gulf Publishing Company
Book Division
Houston, London, Paris, Tokyo

Technology Transfer—
A Realistic Approach

Library of Congress Catalog Card Number 78-62608

ISBN 0-87201-822-9

Published in France as *Réalités du Transfert de Technologie*, Masson, Editeur, Paris, 1976.

Contents

Foreword

The central importance of technology transfer between rich and poor worlds is widely recognized. Indeed it is the only practical way in which the imbalance of wealth can be redressed. Unfortunately, this recognition is matched by astonishing ignorance about the factors that make technology transfer effective. Politicians speak of aid programs mainly in financial terms and industrialists are preoccupied with machines, blue-prints, and hardware.

Silvère Seurat, more than any other individual of our generation, has taught us that money and machines are sterile without the capacity to understand, use productively, and organically develop technological knowledge and skills. Education and training, for him, must always be related to concrete results and not thought of in isolation. He insists on a profound understanding of principles as well as practice with the ultimate goal of "industrial mastery" for the individual, the organization, and the nation.

This book is a storehouse of wisdom, built up by Mr. Seurat over the past twenty years in forty-five countries ranging from Indonesia and Algeria to the United States and France. Virtually every industrial sector has been covered. As a professional management consultant I have been particularly impressed by the interplay between theory and practice. From basic principles and tentative hypotheses Seurat went out into the field and worked with his clients on tough projects: a vast mine to be opened in Africa with no local reservoir of relevant knowledge and skills; the erection and running of a complex petrochemical process plant in the Middle East; helping over a million peasants to acquire new behavior patterns in rice planting so that the "green revolution" could move from dream to reality. Each case called for multidisciplinary contributions—the engineer, the planner, the psychologist. All had their roles as members of the Seurat team. The lessons from each case were carefully studied, analyzed, and codified. So the cycle was repeated until an original scientific method for technology transfer evolved.

The following are some of the features that I have found particularly impressive in this method:

1. The insight that technology transfer above all is the capacity to store and transmit to people the accumulated experience and understanding of others.

2. The meticulous analysis of every facet of knowledge, skills, and perception that make for success in any task and operation. Then the use of advanced methods of motivation and reinforcement to transfer the knowledge and skills.
3. Sensitive adjustment to different cultural and work values.
4. Accelerated development of local managers and instructors so that dependence on expatriates can be reduced.
5. Close links with the educational systems of a community. Indeed, Seurat's method has itself been used with striking success to improve national systems of technical education.
6. Early planning of human and managerial aspects of new projects so that expensive and wasteful "crash programs" are avoided.
7. A systems approach that embraces all aspects of task analysis, organization design, recruitment, training, control, feedback, and dynamic development.
8. A proven capacity to deal with very large scale projects. The needs of the poor world are mind-boggling in their sheer numbers and urgency.
9. A dedication to the most cost-effective way to get results on every project.

This book is essential reading for every manager in international companies who has the responsibility for designing, managing, and bidding for new plants and projects in the developing world. In turn, managers, politicians, and administrators in the developing world will be stimulated and counseled to ensure that plans for the human and managerial aspects of a project are expressed as clearly as those for finance, machines, and materials.

Peter Drucker once said there are no undeveloped countries, only undermanaged countries. Silvère Seurat and his colleagues, now numbering several hundred, have made a practical and original response to this challenge.

John Humble

Introduction

For Whom Is This Book Written?

All nations, whether sender or receiver—and often both—are today keenly watching the means by which industrial knowhow can be exchanged—a process more commonly known as technology transfer.

And in many different quarters, including ministerial offices, company directors, public and private bodies, and engineering companies, the same elementary questions are being asked, *what, when, how,* and *who*.

It was to reply to those questions, so often heard from executives and administrators at every level, that this book was written. Moreover, the author hopes that the following pages will deepen the interest of students and scholars in this new exciting and challenging frontier of human activity.

How Is It Presented?

The book sets out neither to popularize the subject, nor to be a technical manual. A popular book on the subject, (and its usefulness is not disputed) would outline in simple language ideas already mastered and tried out in the laboratory and in industry. But that is not the case as far as technology transfer is concerned. That is why this book, more of a thesis than a synthesis, will not satisfy the hurried reader, even though it does not call on very complex ideas.

The book conveys the actual level of complexity of technology transfer, which is different from the facile conclusion, "Ah yes, so the export of training is what it's all about." It will not try to describe how a professional mastery of the subject can be attained.

What Is Its Object?

We live in an epoch where man, more than machines, hinders economic development. This book contributes to the development of man's expression, techniques, and methods of technology transfer. It is designed to propel an ambiguous art from the age of astrology to the age of astronomy.

What Solutions Are Proposed?

Firstly, it proposes a clarification of language and a listing of the main actors and their roles in a discipline where even the meaning of words is vague and the studies of experts hardly understood.

Secondly, it unfolds a method created through experience of preparing work teams for large industrial plants, and also from the creation of new, specialized training schools. The application of this method is generalized in the case of an entire company and is aimed at fulfilling the company's industrial capability.

Finally, it lays down the broad stages of the method's development as well as ways in which it can be extended and the validity of its field of application.

1

Technology Transfer Is a Human Characteristic

Economics and industrial development, austere subjects though they are, do not escape the tinge of modishness that so characterizes our age. And the result is that an innovative process or formula can suddenly be thrust into the limelight of an international stage more for the way it is expressed than for its content.

Thus the phrase "technology transfer" expresses a fact as old as man and describes a characteristic human property: the capacity to store and transmit to people the accumulated experience of others.

Bathed in wide-spread acclaim, frequently mentioned in an international context, and the chosen theme of numerous top-management seminars, the label technology transfer suffers the inevitable reverse effect of a too-rapid and brilliant ascent—alas, its meaning is clouded by ambiguity.

Technology transfer should be properly restricted to the flourishing and long-established sale of licenses, say some. Yet, others reply that such a course ignores reality in all its complexity, and that to simply place a legal limit on a phenomenon that touches the vital needs of each nation and a large part of mankind is to ignore the complexity of the real situation. No doubt a license is often the origin of a technology transfer, but it is no more than the visible part of an iceberg. The hidden mass breaks down into an intricate chain of operations—organization, personnel selection, education—which are the only means by which a specific technology can be implanted.[1]

We prefer to consider technology transfer with all of its complexity. Such an approach is more relevant to our clients, who are more interested in concrete facts, in making a plant run rather than in the minor details of the industrial property transferred.

---

[1]This point is similar to that put forward by Edward P. Hawthorne in "Technology Transfer," published after a conference on this theme at Istanbul, organized by the OECD in October 1970 (p. 8): "Technology must be understood as the application of knowledge in the whole of the production process, from research to sales. The key elements can therefore be of a different nature ranging from scientific theories to sales techniques at board level."

For as long as technologies have existed they have been moved around within countries, between different segments of the population, and between countries. My company has been transferring technology since 1960 through the introduction of manpower training in new work techniques and improving efficiency within properly structured working groups, set up to do a specific task.

But what's so new about manpower training? Why add one more bit of experience to what is an already well researched topic? To explain that we must take one step further, to demonstrate our experience and to reveal why it is so original.

Industrial reconstruction reached its high point in France and Europe around 1960; in the Third World, industrialization was stirring.

New kinds of plants and factories were being built in France which already gave some hint of the sophisticated and giant machines being developed today. And as they were often built on green fields, they provoked a stunning culture shock for agrarian communities, as if they had arrived from another planet.

In the developing countries, meanwhile, a vast crevasse yawns between workers and the sophisticated jobs that will be offered to them. Giant industries impose their own laws. The same industrial complexity that was once so astounding in Europe is now found in the Sahara and Gabon oil wells, and the mining industries of Mauritania, Togoland, or the Niger.

Yet twin hazards loom ahead: a political risk if the factory jobs are reserved for skilled men imported with the machinery, and equipment deterioration if unskilled local labor is used.

Such hazards are far from imaginary. There are too many actual examples. Avoiding such disasters has been our object for the last 15 years, while being conscious of the use of our pioneering work, and confusedly sensing that, beyond the thankless daily tasks, we were participating in the birth of a new profession that still seeks a name. (We shall name it in following pages "technology transfer consultancy" or "technology transfer enginering.")

But there are no easy answers to the problem of how to create a manpower team for a new plant. Whatever the training or experience of each man, they lack experience of working as a group. A group of dancers does not become a *corps de ballet* overnight; nor a number of sailors and officers the highly efficient crew of a submarine. By the same token the right dose of workers, foremen and engineers, no matter their experience, does not constitute the efficient operation of a new plant. What the drawing together of those workers needs is training as a team: the details of the new plant and its innovative features must be underlined. Each team member must appreciate the key-point in his job as well as in his relations with the other team members. An industrial work force is pulled together only after patient, intensive training, which should be begun before the plant starts up. Such training, more or less vital depending on the level of industrial experience of the work crews and the complexity of the plant, should follow certain general rules, the stating of which is a natural aim for the technician.

On the other hand a plant project is often planned in detail months, if not years, before its completion. And if it is precise in its technical details, it must also be precise in the description of tasks necessary for its operation, and in the grouping of these tasks in jobs adapted to the backgrounds and motivation of the future workers. The goal is not to prepare the workers for an uncertain future, which is the thankless and difficult task of any state educational system. On the contrary—and this asset is priceless—its aim is to train men and women for a certain future some years hence; that is their definite job, perhaps even the beginning of a career, in this newly-born plant.

A plant under construction is a fruitful teaching opportunity, offering a natural synergy, and echo between training and action. Used properly, and going beyond the technical details to the overall professional behavior, training sometimes means following in a few years a route that might have spanned decades or even centuries. Thus, since 1960, we have been training factory and plant personnel through Eurequip, a company which is specialized in this field and which will often be referred to in this book.

A second closely defined operation has also occupied us since 1960: meeting the wide demand for the training of personnel in similar administrative or factory jobs.

There again, the route between training and its implementation is even easier with the training of teams for the new plant. But we quickly realized that all-embracing, general solutions were no good. There was a need for original training programs adapted to the background of the trainees and the exact nature of their jobs.

That second specialization has become even more important than originally realized. A technology transfer situation often involves a large number of men and women who will be called on to perform different tasks. The operation breaks down into several "man/role" segments which group together the whole team embracing anything from a handful to several hundreds. The result is that the basic training system of a homogenous "man/role" group underpins all technology transfer systems.

That is also particularly the case with the preparation of industrial teams. The solution involves the defining of a certain number of groups made up of workers who are to receive the same training. Our two starting points are thus complementary: preparation of an industrial work force and creation of an elementary training system.

Lately, our twin tasks have been enlarged. We have now progressed beyond the "man/role" group and the factory to the whole company. In fact, a new plant is nothing more than a particular example of the continual changes the company undergoes and to which people must adopt and initiate others. So, from training factory work groups we have passed on to the preparation of management teams for major company operations including research and development, marketing, data compilation, purchasing, personnel, finance, environment, planning, and even general management. Our methods of training these men and women in new methods or new attitudes to change differ little between a country with a long industrial tradition and one which is still developing.

The company, "the university of life," in every case reveals itself as the pinnacle of technology transfer, pulling all its members along in its wake in a joint dialectic between know-how and action, permanently tying together individual and group performances.

Secondly, we have moved on from the preparation of industrial teams to the idea of indispensable pretraining systems. These are conceived without reference to other existing systems in other countries, but solely in relation to the stage of development and sociological characteristics of the men in the country concerned. Not only is the training system a vital element in the development plan, it is also linked with the development of each individual and prepares him for a role in the economy of his country equal to his ability.

For there is no successful route to technology transfer, and thus the economic development of a country, beyond the patient, detailed forging of a chain between the individual to the nation, linking each workshop, village, factory, service, administration, research center, and specialized company, whether it be public or private.

The theme of this book is not the whole panoply of technology transfer, but solely that aspect we have lived through during the last 15 years—with all its difficulties, fumbling, and mistakes.

The following pages take their matter from the world of action. The message is both incomplete and biased. Incomplete because it reveals only our own work, biased because it exposes simply the operational methods that we have developed in the field.

The plan is simple. Chapter 1 gives an example of a technology transfer of several years ago, Chapter 2 introduces some general ideas, Chapter 3 attempts to condense our experience into a method, Chapter 4 is devoted to the setting up of a training system before the transfer operation, Chapter 5 details the origin of our methods already lightly sketched above, and Chapter 6 sums up and contains certain thoughts on the conditions for real progress in technology transfer. The appendix contains seven chronologically ordered texts published between 1960 and 1975 dealing with an application or a point of method.

All that remains is the agreeable task of paying tribute to everyone to whom it is due, while emphasizing that this book is not solely the author's own work.

Its roots are in the accumulated work of the members of our company, and represent in total a thousand working years, summed up in the knowledge and title "engineer."

In several passages throughout the book, certain ideas and even texts, are the work not of the author but of a past or present member of Eurequip. Their name and contribution are each time attributed. Special thanks are given to Pierre Revol and Françoise Roquigny for their constructive cooperation during the preparation of the book.

And finally, if we had not found courageous clients ready to put their trust in our frail teams, none of which is narrated would have happened.

2

A Case History[2]

In the South Sahara desert is found a branch of an industrial sector which is in the forefront of technological progress: a uranium mine and its treatment plant which are preparing for start-up.

Background

The Société des Mines de L'Aïr (Somaïr) works a uranium ore deposit located at Arlit, on the southern border of the Sahara, in the Republic of Niger. The deposit was discovered in 1966 by the French Atomic Energy Commission following extensive prospecting in the vast sedimentary basin stretching westwards from the Aïr mountains.

The company was founded on February 1, 1968. Today the shareholders include the Atomic Energy Commission, the Republic of Niger, the Pochiney-Mokta mining company, the Compagnie Française des Minerais d'Uranium, the Urangesellschaft, and Agip Nucléaire.

The operations consist of open-cast mining followed by chemical processing of the ore. Processing consists of crushing the ore to 800 microns followed by sulfuric acid leaching. The heavy liquid recovered by washing is next filtered and purified. Lastly, uranium is precipitated out in the form of sodium uranate titering about 70%.

The plant came on-stream in January 1971; it can process 350,000 tons of ore per annum, producing 750 tons of uranium in concentrate form. Capacity doubled by 1975.

The design and construction of the project posed certain problems because of the location of the deposit. Arlit is located 2,000 km from the sea, in a practically uninhabited region. Moreover, the Republic of Niger still has a very low level of industrialization. Somaïr therefore had to create not only the required

[2] Extract from *Travail et Methodes* no 283, November 1972; written by the management team drawn from SOMAÏR and Eurequip (Pierre de Boisanger and Jean Jacques Saint-Andre).

industrial infrastructure but also a mining city including housing for the personnel, leisure centers and playgrounds, a school, and a hospital.

A further problem was the still very limited resources of Niger in terms of trained personnel. Only 10% of the population received any schooling. The recruitment and training of the personnel was therefore one of the company's primary concerns.

The same training approach was used for all the personnel, whether for the mine or for the processing plant. Nevertheless, special problems were posed for the plant; which is why we have chosen it as an example to describe the methods used by Somaïr.

Preparing the Training Plan

It was first necessary to formulate objectives, define the jobs to be done, and assess the available human resources.

Economic and political considerations favored a policy of using local manpower as much as possible, while laying plans right from the start for the gradual Nigerization of as many jobs as possible. Further, the geographical isolation of the plant required that it be virtually self-sufficient in the matter of maintenance; hence it was necessary to have staff available on the spot who were trained in a number of specialities.

The engineering design of the plant, the detailed design of the jobs, and the human resources survey could not be done independently of each other, because of the impact each would have on the other. Close cooperation was needed to find an optimum solution between the use of local manpower and the use of highly automated equipment, which required well-trained, specialized workers; i.e. expatriates, particularly for maintenance.

These different tasks were successfully accomplished through the assistance given to Somaïr by Ugine-Kuhlmann, responsible for the industrial design of the plant, and Eurequip as consultant for the training of the personnel. A general training plan was drawn up in mid-1968, and was followed in greater detail as the design study progressed.

The Overall Approach

Figure 2-1 presents the overall process used for the training program. For ease of comprehension we have omitted references to the many links between the training program, the industrial design process, and the design of work methods. Also, the lengths of the lines in Figure 2-1 are not proportional to the time required for the operations.

The following are some of the main features of the training program:

1. "Cascade Training"—a method in which the managers and engineers prepare the training for, then train the supervisors and technicians, who in

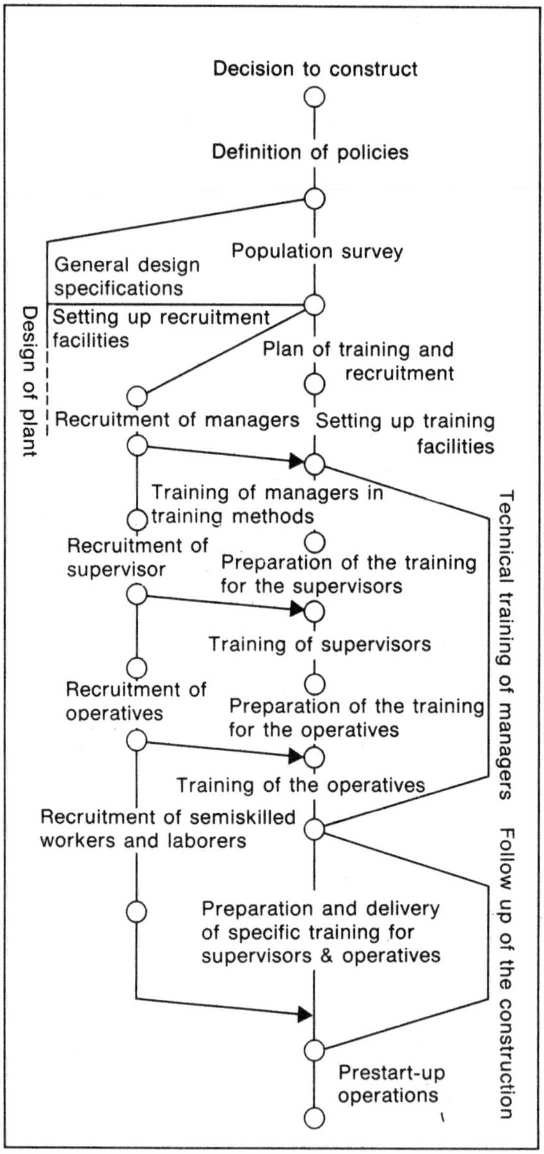

Figure 2-1. This diagram illustrates the overall process of preparing the operating crew.

turn prepare and conduct the training of the operators, supervised by the managers. This presupposes the setting up in advance of a training organization and the elaboration of training procedures. It also requires that the operating manuals for the plant be ready.

2. The tasks of organization and training—that is, the design of the jobs, the organization of the work posts, and the preparation of training courses designed to cope with the starting level of the trainees were carried out by the plant's heirarchy, which has been trained for this role and for these responsibilities.
3. The selection and recruitment were phased to keep pace with the training schedule and took account of expected time-lags in the response of the labor market. These also required the setting up in advance of a recruiting organization and the establishment of selection and recruitment procedures.
4. The technical training of the managers and supervisors was conducted in parallel with their activities as trainers and was enriched by these activities. This training also included the progress and follow-up of the erection of the plant and involvement in the start-up operations.

The schedule shown in Figure 2-2 gives the approximate duration of the various phases in this "Cascade Training" program, based on the total involvement of managers and supervisors.

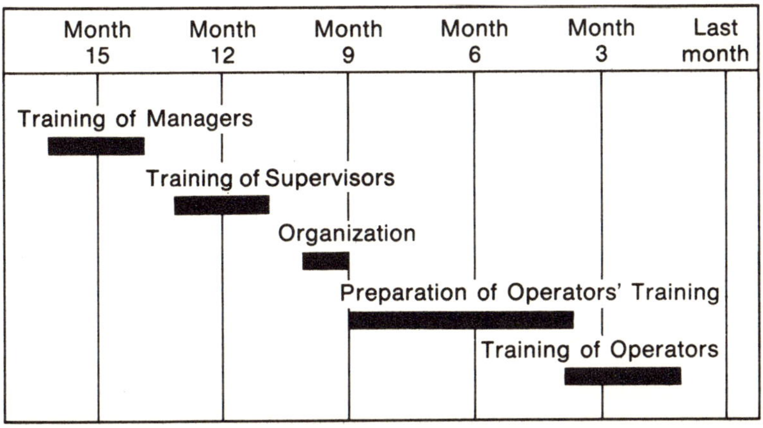

Figure 2-2. These are the approximate durations of the various phases in a "cascade" training program.

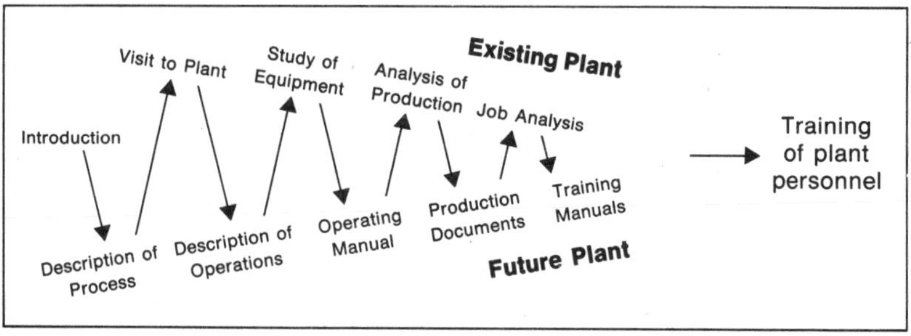

Figure 2-3. The training of managers and supervisors was based on a two-way process: the observation of the "reality" of a plant and the preparation of key documents concerning the future plant.

Training Methods

Training the Managers and Supervisors

The training of the managers and supervisors (shown in Figure 2-3) was based on a "two-way" process: the observation of the "reality" of existing uranium treatment plants and the preparation of the key documents concerning the future plant.

The first step, the introduction, gave general information on living conditions in Niger and Arlit, on the production process, and on the organization of Somaïr. The trainees were then introduced to the active training methods (or self-training methods) that were to be used: preparation of the key documents, observation and analysis of jobs, interviewing techniques, etc.

They then had to organize their own departments or sections, their own jobs and those of their subordinates, and establish the liaisons with other departments upstream and downstream in the production process. This included hierarchical liaisons and relations with staff departments. After this they designed, prepared, and delivered the training courses for the operating personnel. These tasks of organization and training were the major elements in the "self-training" of the managers and supervisors and in their integration into the company.

To prepare them for their training role, Eurequip gave a two-week course in training methods, including: (1) the design of a training course based on the difference between the starting level of the trainees and the level required for the job, and (2) the preparation of theoretical and practical training sessions, and of the design of appropriate teaching aids (audio-visual, models, etc.).

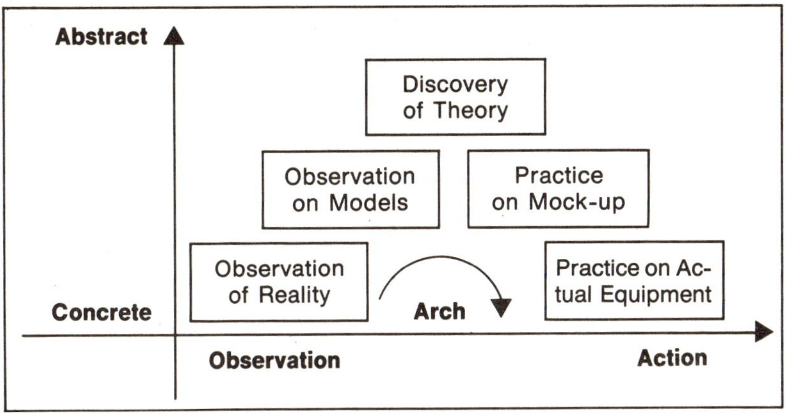

Figure 2-4. The arch training method is based upon the principle of active training in small groups.

Training the Operatives

This training was based on the principle of active training in small groups. Each session was organized according to the "arch" procedure comprising the following successive phases: (see Figures 2-4):

1. Observation of reality, i.e. the equipment in the course of erection, installation or testing
2. Observation of a "model" of reality, which was more comprehensible and accessible
3. Discussion of the model to acquire a thorough theoretical understanding
4. Application of this newly-acquired knowledge on a mock-up of the "real thing," and practice on this mock-up
5. Further practice, but this time on the real equipment

In a plant under construction the "real equipment" is often inaccessible for obvious reasons, which is why observation and training on simulators or mock-ups has understandable advantages.

Visual aids were used representing the individual components of the work post, the main and secondary circuits of materials, and control systems.

The elements were mounted and assembled on magnetic boards at each session. The assembly of the parts already studied provided an opportunity at the start of each session for checking and reinforcing the previously acquired knowledge.

Table 2-1

Personnel Assignments

12 Heads of processing units	10 Laboratory assistants	13 Maintenance mechanics
48 Processing operators 4 Heads of contact units	6 Samplers 5 Control and regulation operators	2 Lubrication men 2 Plant electricians
5 Contact operators		8 Plant welders

Since the processing is mainly automated, control of the entire plant is carried out from four control rooms. Four wood and cardboard full-scale control room simulators were built to recreate these control operations in both normal and abnormal situations.

Operators were trained mainly on these simulators. Instructors created failure situations and the trainees had to simulate the necessary sequence of operations to return the facilities to normal. At the end of the training period the managers, supervisors, and operators participated in the equipment trials alongside the start-up crews.

Results

The recruitment operations led to 155 candidates being selected for training including 36 of the B.E.P.C. level (approximately junior high school level), 77 of the C.E.P. level (approximately high school level), and 42 of the C.A.P. level (technical training). Of the total, 115 were kept at the end of the training period for the operation and maintenance of the plant. Their assignments when the plant started up are shown in Table 2-1.

On the whole the results were very satisfactory. Less than six months after start-up, the personnel was able to run the plant without outside assistance. For the maintenance staff however, the results were not so immediate: training maintenance workers from a population without any prior industrial background is necessarily a long process. Further, the practical training of such operators requires that the relevant plant and equipment to be available, which was not the case during the construction phase.

In addition, the training method employed had the following results:

1. Each manager and supervisor was highly trained in relevant techniques and management responsibilities.

2. The operating teams were formed around the twin relations trainer/trainees and superior/subordinates.
3. The training organization and the methods, procedures, and material facilities for active teaching could be used for subsequent phases of training, aimed at giving workers a second necessary skill or preparing them for a new job.

The overall cost of the training for the plant may be estimated at $4,500 per successful trainee (i.e. assigned a job in the plant). The cost includes

1. Trainees' wages and overhead
2. Instructors' wages and overheads
3. Consultants' fees
4. Expenditure on material facilities: premises, furniture, teaching aids, simulators, etc.

Put another way, the costs of training the personnel amounted to 3% of the plant's capital investment.

Conclusion

Somaïr achieved its essential objectives of

1. Aiding Niger's technical build up.
2. Having local personnel available within the required time limits and for the maximum number of jobs, who would be capable of operating and maintaining the plant.
3. Avoiding the direct and indirect costs of employing expatriate personnel to man the positions.
4. Avoiding costly problems during the start-up period resulting from inadequate training of the operators, which could have aggravated other problems which arise in the assembly and testing of new equipment and a new advanced method.

The special conditions of the Arlit plant made it particularly necessary for Somaïr to undertake this effort for the training of the personnel. It is important to emphasize, however, that experience showed that such an operation was profitable. It is probable that this approach to the start-up of a new plant could be profitable even in industrialized countries.

This case history, begun in 1967 and written up in 1972, has now lost its adventurous aspects and become an industrial reality. Not only has the plant been running at its nominal capacity for some time, but extensions have already been built and others are planned. The board can delegate day-to-day running problems to management and is concerned only with long-term, strategic problems. Let us quickly analyze the Somaïr case by replying to the key questions, *what, when, who,* and *how.*

Was There a Technology Transfer?

There are two answers to that question; the first is immediate and affirmative; the second more thought-out and subtle.

The first reply is immediate and affirmative because workers far removed from a technical environment are now capable of using, in teams, modern mining equipment and running the various uranium treatment systems throughout the plant. All these groups master, in their areas, the jobs linked to the technologies used and work together in a perfect harmony comparable to anything found in the best-run factories and plants in industrialized countries.

Yet, the answer is more profound and subtle if we add that the labor force is not totally from Niger and that the groups involved in the transfer contain expatriates in the management and technical jobs.

From that appears a general characteristic of the transfer—its duration. The transfer began before start-up; it has been followed up since then and will continue to be followed up because follow-up is included in the structure and working relationships in the plant. It is both a production and a technology transfer system and everyone is well aware of that.

When Did It Occur?

The transfer is never really completed. One can always progress down the road of technical mastery. Preparation for start-up was spread over four years. The young, mixed, operational teams reached their first objective in correctly operating the equipment. They then aimed at, and reached, more ambitious targets. Thus, the transfer will progress following a plan that takes into consideration workers' and groups' evolution, preparing for and facilitating the increasingly confident mastery of professional situations. Management forecasts, worked out during the preliminary study in order to construct the training plan, are based on increasing integration of a Niger labor force and reduction in the number of expatriates.

Who Were the Actors Involved in the Transfer?

The presence of various bodies has already been raised. On the one hand are the companies or plants preparing the perfect mastery of the technologies to be transferred, the technical engineering methods of process modification; on the other is the receiver and his environment, ready to receive and undertake the new technologies. Between these two is the consultant, undertaking the vital liaison; he conceives and implements the transfer.

How Is the Transfer Achieved?

Methods are discussed; an approach is succintly described. Each step is important, but the essential starting point is the creation of a transfer strategy based on the in-depth study of technical, structural, and cultural environments involved. The differences that come to light must be the object of management modifications and changes in the jobs so as to aid the transfer. The mixed work groups of well-selected individuals are carefully prepared to be transfer leaders. Training is specially prepared for that problem, adapted to the trainees' level and the jobs forecast.

We shall stop our analysis of this typical example and try to look deeper into the questions surrounding technology transfer.

3

Definition, Main Actors, and Dimension of Technology Transfer

In Search of a Definition

The Somaïr case study has familiarized us with the facts of a technology transfer. The work of Edward P. Hawthorne, already mentioned, opened the way towards a general definition of the concept. It can be detected in the expectation expressed by companies demanding transfers of technology. A demand for a turnkey and products-in-hand contract, despite some confusion, offers a clear expression of these expectations, "obtain an effective control of the functions involved in a technology."

That leads us to the following definition:

> *A technology transfer takes place when a group of people, usually belonging to one body, become capable of performing one or several functions attached to a specific technique in satisfactory conditions.*

Starting from this definition—formulated explicitly so as to define a complex subject and establish communication with the reader—it seems a good idea to select certain terms demanding explanation.

Technology Transfer

The term technology transfer deserves to be retained as much for its success as a description as for its conciseness and clarity.

By transfer we mean "the moving of something from one person to another." Two essential ideas emerge from that. Following the idea of something being transported, there must be a departure point and a destination. There is no sense in the receiver of technology reinventing what already exists. Rather he should benefit from agreements with the owners of technology. There is thus always a sender who owns technology and a receiver who wishes to take it.

But take what? A technology is the explanation of a technique, otherwise expressed as the knowledge related to a technique. Yet often in fact, rather than knowledge, it is a question of the professional written or nonwritten know-how of a group, or of individuals, which allows others to master the technique. Probably the term "transfer of technical capability" would be better still, so that technology transfer retained its conciseness.

One last remark about the term technology transfer must be made—notwithstanding the aforementioned definition—the complexity of a technology transfer operation can vary between a simple grafting of a much-copied model, to the transfer of a completely new and original system. To some extent it is a pity that the phrase tends to oversimplify the process, but words cannot convey every nuance. But let us not forget that the simple transfer of a technology is a different matter from its adaptation to alien socio-economic conditions.

Technique

The term technique, already mentioned in the context of technology, is the last part of our definition. It is a key word with two meanings: one narrow, the other larger. In a narrow sense technique is, "a number of processes based on scientific know-how used in production." That meaning is the one most frequently understood when talking of technology transfer.

But in fact we have already seen that this definition is too restricted; industry and the economy are not based solely on production and its techniques. Other techniques, sales, management, finance, accounting, planning, and so on, are necessary too.

So a wider definition must be sought. The dictionary also gives us "a number of procedures used to produce something or to obtain a previously determined result." That second definition opens the way to two types of transfer according to whether the procedures involved in the technique apply only to workers or to machines. In the first case we more easily understand that "techniques of management" are referred to; in the second case "production, maintenance, machine tuning, or research" are implied.

But there are striking differences between the two types of transfer. In our second case, for example, the new techniques are ushered in by new machinery that will be housed in a vast factory or plant, the construction of which itself will make a resounding impact on the region or country. While such new machinery will disturb the local population, it will also attract them, and learning how to master machines can become both an introduction and motivation for the local population. Such motivation will be found less easily when it comes to management of such jobs as parts inventories, preparing of orders, personnel, wages, and so on.

But complex machinery demands rigorous attention. And that renders transfer even more difficult; it will not stand the trial and error methods that are acceptable in management.

Working Groups

In industry all work is carried out in groups. Of course, one person can initiate a vast enterprise, but it is people working there that give it importance and meaning.

A transfer can be the work of one man who draws together a team around him. But the operation cannot be judged a success until the whole team has learned the necessary skills. The composition of the group may also vary according to the nature of the transfer. The Somaïr case demonstrated that frequently a transfer success relies on the presence in the transfer group of expatriate executives who enter at a high level in the hierarchy. As the transfer progresses, the expatriates one by one leave the group to be replaced by the best men on the spot. Thus, the percentage of local labor is the first qualifying factor in a transfer.

But if we associate the idea of a transfer with a group, we must leave out certain manual crafts—even though such crafts can be an essential element in a country's prosperity and its capacity to receive sophisticated technology.

Excluding such crafts is not to pass an unfavorable judgment on them. It is another kind of problem, with deep, local, cultural roots. For while industrial techniques pay no heed to national frontiers, craftsmanship draws its strengths from traditions that must often be rediscovered before they in turn can be transferred.

Craftsmanship is thus on the fringe of technology transfer and will not be referred to again in the following pages. That does not mean that the methods we propose cannot be applied to crafts, but rather that it has so rarely arisen as not to interest us.

Usually Attached to an Organization

The group of people involved in the transfer cannot be a temporary group. It must belong to an economic body which, because of its permanent structure, is the only one capable of effecting the new technology. Often, it will be a private or public company, or one with both state and private capital. Sometimes the sender of the technology will take capital participation. But the organization may equally be a research center or school.

For simplicity's sake we normally refer to the technology receiver as a company, though in fact it may apply to other kinds of organizations.

Satisfactory Transfer Conditions

There is no such thing as a perfect transfer. There are only relative transfers in the sense that a group of people who once could not run a plant, now can. That means that they run the plant with a certain level of production while machine

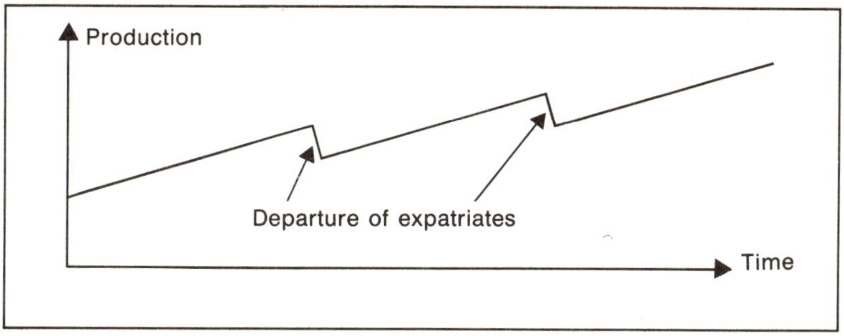

Figure 3-1. This diagram illustrates how the departure of expatriates (members of the control company) affects production.

break-downs and waste material do not rise beyond a certain level. Yet these performance levels, which must be fixed beforehand, are only indications. Another, better-trained group and even the beginners' group after several months training, would do better.

When we say for example that X knows how to play the piano, we are making a relative judgment. We mean X can reach a certain minimum standard. But we know equally well that Mozart and Beethoven played better than X and that X will get better with more practice.

Thus, a second qualifying factor enters the transfer, related to the performance of the plant team.

There is a link between the two qualifying elements of local labor and production. It is that as production goes up, certain expatriates can leave. But that will be followed by a slight drop in production to be made up several months later. More expatriates leave, production drops, then picks up, and so on. This process is illustrated in Figure 3-1.

One or Several Functions

Grouped around a technique (which may be embodied in a machine) are certain jobs that often must be carried out by different groups of workers. Such jobs require differing degrees of personal initiative—and thus to a third progression concerning the running, maintenance, setting, improvement, assembly, or even invention, of that machine.

Where Does the Transfer Come In?

The transfer can take place either within a firm, within a country, or from one or more countries toward one or more countries.

Table 3-1
Various Technology Transfer Situations

Transfer Areas / Systems Involved	Within the Same Company		From One or More Companies to Another	
	Isolated Individuals A	**Groups B**	**Within a Country C**	**Within Different Countries D**
Men (management)	Example: reception of a recent graduate, and reinsertion of an experienced manager following intensive management training	Example: introduction of strategic management	Example: orienting a university research center to industry	Example: introducing American management methods in Europe (the 1950s and 60s)
Men and machines (production, maintenance, setting, research and development, data processing)	Example: orienting a surveyor to an instrument control setting section	Example: applying a research—discovered technology to manufacturing	Example: introducing builders to future users of new factories—NASA to American industry	Example: German space technology to the USA and Russia (around 1945)

Technology transfer as we have already noted in the introduction is a widespread phenomenon, a characteristic of man and development.

The transfer can take various forms. We note four, lettered *A, B, C,* and *D, of* which *D* is the most frequently understood. A condensed version of these forms of transfer is contained in Table 3-1. Each can apply to a different management system, or man-machine system.

Situation A

It concerns essentially the individual in a company. It could be for example the entry into a plant of a young manager or worker who has just graduated from school. In the best case, the transfer works in two directions: from the company to the individual, then back again from individual to company.

From the company to the individual. The recent graduate steps into the company with a theoretical rather than practical background. This theoretical knowledge should be applied if it is to be used to its best advantage. If the company is sufficiently well staffed by individuals of all levels, who understand both theory and practice and whose number goes beyond a certain "critical mass," the transfer will take place naturally. The beginner will receive the best kind of introduction: one in which he does not give up his theoretical knowledge but uses it to solve practical problems.

But in many cases the "critical mass" does not exist. A foggy pragmatism governs all work with little reference to a scientific approach. In this case, the beginner wants to integrate, he will be forced to cast off his theoretical knowledge, which he will little-by-little forget anyway, and stick to the empirical rules laid down in the company. As a result, the "culture" of the beginner and the "culture" of the company will not graft together if the gap between them is too wide. However, above a certain level the graft takes place at the lowest common culture denominator.

From the individual to the company. If the initial transfer is properly carried out there will be a fruitful communication between the beginner now integrated and the company. He may in turn enrich the working environment with his theoretical knowledge. And even in this example, which appears to be centered on the individual, the successful transfer involves individual and group, with each member being more or less transformed according to the degree of integration.

The same situation applies to a qualified manager who has received intensive management training outside the company. If the training was at a higher level than the management culture of the company there is a good chance that the grafting will not take. The manager will rapidly forget all he has learned and which is so different from his actual working environment. All of that training was wasted, it was merely "water off a duck's back."

Finally, there is the fruitful transfer of someone from a highly specialized technical department into a production-orientated one. A measurements specialist for example, might arrive in a shop for the calibration of control equipment. Assuming the graft takes, the section will learn and understand exactly how measurements should be made and how to apply it to a production line.

Situation B

This situation concerns related groups within a company. For example, the progress of a certain technology from research and development to manufacturing and even sales and after-sales service. It can also refer to the introduction of certain technology into a company which improves performance at all levels, such as data processing or management planning. With the technology arriving from outside it is a situation similar in some ways to *C* and *D,* and deserves much greater attention than it usually gets.

It would be delusive to present a company as though it were a community of minds in which the knowledge of one could be freely transferred to another who could translate it into action. The company is a microcosm containing different cultures and consequently the same kind of communication blocks that exist between countries.

Those differences present themselves as a Babel of jargons and value judgments among men who are separated by their education and membership in closed groups. They are the communication gaps that exist between the manufacturing and maintenance specialists, between research and production, marketing and supply, accounting, and many other services.

These almost permanent difficulties are particularly noticeable during the introduction of a new technology and have been the object of numerous commentaries, whether they involved economic calculation, data processing, planning or, more recently, ecology.

Pierre Davous (deputy general manager of Eurequip) and I have tried to characterize by different methods the way in which technology is introduced into companies.

Pierre Davous, who is always aware of the need for concrete results insofar as they involve companies, distinguishes three:

1. *The differentiation stage*. A specific "cell" is created to which specialists in the new technique belong, among them some from outside the company.
2. *The integration stage*. The specialists, while remaining in their cell, establish a common language with the company's managers. They gradually introduce the new technique into operational and strategic decision making, while also cautiously increasing its complexity.
3. *The general acceptance stage*. The new technique becomes available to everyone—as commonplace as the telephone.

For my part—because I am anxious to show the changes in the company which lead to new actions—I propose to distinguish three kinds:

1. *The anatomical change in the structure*. It is decided to create a new cell; then later eliminate it.
2. *The cultural change that permits communication*. The specialists in the new cell train the managers at every level in the new technique following upward training paths.
3. *The psychological change in the structure*. The new technique finds its place in the arsenal of methods used by the company and gradually widens its scope by a double movement, on the one hand by a gradual spreading movement, on the other by increasingly complex new situations and tools.

Table 3-2

Structural and Cultural Changes in a Company Caused by New Technology

Behavior Stages / Nature of Changes	Differentiation	Integration	Acceptance
Anatomical	Creation of a new specialized department; establishment of a small cell of personnel (1)		(5) The specialized service department is cut back; the new technique, now a tool used by everyone, is taken over by all managers
Cultural	(2) Familiarization of that personnel with company problems	(3) Transmission of the technique's potential to management by the new personnel	(6) Acceptance boosts the technique's spread
Psychological		(4) First uses of technique by management are helped by the new technique's service department	(7) Double growth in use—in width and depth

Table 3-2 sets out the structural and cultural changes in the company with regard to the modifications brought about by the new technology. This is not the only route they can take and a shorter path is possible.

What is important is to be well aware that the technology transfer does not simply affect individuals. To equate the evolution of a whole group with a simple training exercise, or to equate technology transfer with the exporting of training, is an absurd simplification that can only lead to grave errors.

Situation C

Within a country the transfer can be from one organization to another. That is what happens, for example, when university research transfers to industry. It is also evident in the relationship between the builder and future user of a new plant before it comes into service. One of the best known illustrations of this situation is the spin-off to American industry from the NASA space program. At a more specialized level there is the relationship between the French Atomic Energy Commission and various French industrial groups involved in the uranium and plutonium cycle.

Situation D

This transfer is between one country, or group of countries, to another country, or group of countries. Such is the present case with countries approaching economic take-off, which need technology for their industrial build-up. But it embraces a much larger situation. Three examples, from relations between Europe and the United States, can be cited.

The first concerns the initial transfer of European nuclear know-how to American industry. That transfer gave rise to the first nuclear reactors in 1942 and the first atomic bomb in 1945, only seven years after the discovery of the fissile reaction of uranium in France and Germany. The cause of that successful transfer was the flight to the United States of atomic scientists seeking asylum from racial or political persecution.

The second concerns the transfer after 1945 of space technology and science from Germany to the USSR and United States.

The third centers on the spread to Europe of American management methods, which so characterized the 1950s and 1960s (IMEDE in Lausanne; CEI, Geneva; INSEAD at Fontainebleu; FNEGE in France...).

On a more prosaic level this situation is repeated in the exchange of licenses between industrialized countries. A close examination of this last example leads to the distinguishing of two variants in the attitude of the licensee.

The licensee may consider the source of the innovation to be fixed. He may be content simply to ensure a satisfactory transfer of technology into his company as far as its implementation is concerned. But he might also use the transferred technology as a stepping stone to technological autonomy. He may exploit the technology, which generally concerns solely manufacturing techniques, in his own research laboratories, trying, by detailed modifications, to escape from the restraints of the license. Such was the case with Germany's Kraft Werke Union (K.W.U.) which began by buying licenses for Westinghouse and General Electric light water reactors in order to later get away from them. Similarly in Japan Saburo Okita notes, "A characteristic of technology imports is its constant improvement."[3] That brings us back to the distinction noted earlier between the many different jobs that can be transferred and their classification by people's ability.

[3]"Causes and Problems of Rapid Growth in Post-War Japan and Their Implications for Newly Developing Economies," by Saburo Okita *(Japan Economic Research Center,* March 1967, p. 13).

A licensee, or preferably the technology receiver, can thus follow a course that will lead successively through all the stages of a technology transfer, from introduction and installation to improvement and invention.

What a Transfer Covers

A technology transfer can be broken down into six key operations:

S—Setting up of a system that reveals the job to be done
O—Organization of individual jobs within both a micro and macrostructure
R—Recruitment from outside and inside the organization of individuals with appropriate skills
T—Training of each person for his or her job
G—Group training in the actual working environment
D—Dynamic of the whole: preparation of careers, jobs, and technical system

The transfer must cover these six operations either by repeating them to the sender in a carbon-copy, or by taking into account the needs of industry and the socio-cultural environment of the sender.

The Diversity of Technical Systems

In the preceding paragraph, dealing with the definition of technology transfer, a certain number of key words were revealed.

First were sender and receiver, departure point and destination of the operation. Following that were words dealing with technical capability, the object of the transfer characterized by know-how implying certain individual and group behavior. The capability is implemented through a technical system (unitary systems also) in which a number of procedures are used to produce something, or to obtain a predetermined result. Such a system can also extend to a vast number of men and machines transforming material, the machines maintained and worked by the men, or be limited to personnel to whom information is sent to be collected or processed.

Before exploring an actual transfer problem, it would be useful to analyze more fully a technical control situation by replying to the two following questions:

1. What are its characteristics?
2. How are they put into effect?

The characteristics are typical of a technical mastery situation (and it must particularly strike the neophyte visiting an industrial site, workshop, control room, laboratory, or administrative department).

What is striking about this situation is the existence of specialized, trained groups in which each worker knows his or her job and is capable of carrying it out alone or with others. Behind this phenomenon of the organized group there are a certain number of worker/job groupings, each composed of an individual well-trained in a specific job.

An examination of such groupings places the emphasis alternately on the person and the job. The job, for example, is a collection of elementary tasks that follow certain criteria; the whole must be compatible with the physical and intellectual abilities of the person. It can be more or less personally enriching and leave a small or large part to personal initiative.

Beyond the job we find tasks that stem from the technical system: and to attain the system's goal the machines, if they exist, must fulfill their functions and the workers must accomplish their tasks. The whole is grouped in the jobs, omitting no task, nor unnecessarily repeating it.

The individual's career path must match his or her training and such a harmonious pairing presupposes a mutual worker/job success.

Increasingly in the following pages, we will clearly show that technical control depends on a high level of organization. But it need not necessarily be expressed in the form of detailed procedures, job studies, training programs, etc.

The group can perform the "quicksilver" job solely from the key elements contained in the group's memories. That is what usually happens when the company has mastered the technique to the point that it becomes commonplace (see Table 3-2).

Determining the Characteristic Elements

In addition to those elements previously examined—the trained group, efficient worker/role groups, well-defined roles, job lists, well-selected, properly trained personnel—other operations must be carried out successfully, sometimes in a detailed way, sometimes completely improvised.

The most outstanding characteristic, the virtuosity of the group, is the result of organized or spontaneous group training in an actual working environment. It is designated by the letter *G* —for group. Besides *G*, two other paths lead: (1) to the worker's role and (2) to the person trained in his or her job.

Exploring the first path uncovers the architecture of the work in which each role is separately conceived for each individual by bringing together simple, well-chosen tasks. The architecture at the level of simple individual tasks defines a microstructure within the organization's macrostructure—its division into main functions. Microstructure and macrostructure are revealed in the program.

The second operation we will call *O*—for organization and organigram.

Before *O* is attained, the technical system must be conceived and, particularly, those tasks defined that are to be carried out within the system. The system can be studied by reference to the general structure, especially if that involves the installation of machinery. What is meant by the operation designated *S* is the part of the worker in the functioning of the whole enterprise in every relevant aspect, which can be selected and listed in detail for research, and, in a more general sense, progress.

The training operation, *T*, cannot be divorced from the job, or the cultural and professional characteristics of the workers to be trained. These workers were chosen during a previous recruiting stage, *R*. Depending on the demands of the job, as well as certain policy decisions, *R* involves choosing the best individual in a certain internal or external environment, and it includes the awareness of the characteristics of the men already engaged, so as to adapt them to operations *O*, *T*, and even *S*.

Are the five enumerated operations enough? Perhaps to start with, but not in the long run. The worker/role and group/structure discussion would be meager if it were not for the fact that the situation is changing both all the time. In practice they improve in the job, acquire a better understanding of the structure while their expectations rise: an enlargement of the role is tantamount to promotion. The enlargement can result from a regrouping of the tasks, which in turn modifies *O;* the microstructure or the appearance of nobler tasks—in other words a more complicated technical system, *S*.

Thus a system of machine maintenance, for example, can be conceived at various levels of complexity, each depending on the workers' skills and the development of their industrial culture. A forecasting operation (almost an organization of the future) is therefore necessary for the dynamism of the whole enterprise: the workers, their roles, and the technical system. It is designated *D*—for dynamism and development.

The six operations laid out here—*S* for system, *O* for organigram, *R* for recruitment, *T* for training, *G* for group, *D* for dynamics of the whole—are the keys to technical control. They are the six keys to technology transfer. Figure 3-2 illustrates their logical sequence, which appears schematically as a "T"—for technology and also transfer.

Operations Systematization

We said earlier that the characteristic elements of technical mastery are always there, but are not always apparent or formulated in texts. And the same applies to the operations leading to mastery's characteristic elements.

In fact the more efficient the group is in controlling its techniques, the more difficult it will be to detect the elements and the operations involved. Watching a *corps de ballet,* it is both easier and more agreeable to admire the performance than to reflect on the enormous amount of work behind it. If we were to watch the beginners' ballet class no revealing transfer would result. And yet the beginners possess a kind of embryonic *corps de ballet*.

Can we define the criteria leading to a greater systematization of the key operations of a transfer?

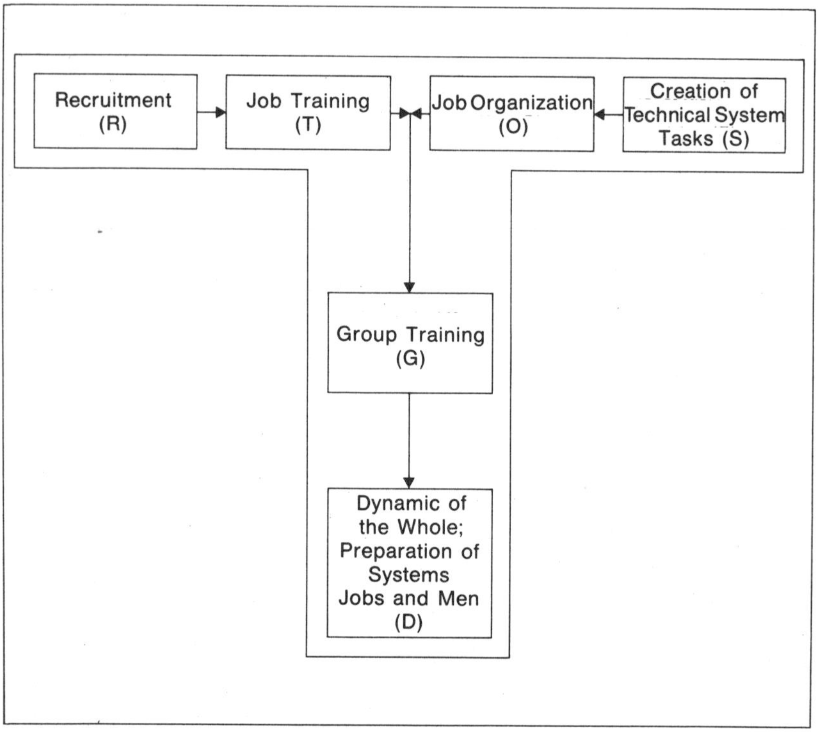

Figure 3-2. Shown here are the six key operations leading to technical competence, i.e., the technology transfer.

First appears the gap between the worker and his or her job. It is easy to understand that the wider the gap the less must be left to chance; all the elements for success must be assembled and studied. On the other hand, if, as is frequently the case in industrialized countries, there exists a rich industrial culture, nothing prevents the operations being less well defined, leaving the working groups some liberty of choice and initiative. Such is the solution adopted in certain autonomous working and project groups.

Next we come across the responsibility of the job, which can be expressed in money, in accident risks, or even in human lives. The greater the responsibility, the fewer opportunities for trial and error and the more detailed must be the control operations.

A third criterion is equally significant: the numeric and geographic dimension of the group united in action. The larger the group, the greater the diversity; organization, education, and group training are more complex, but are all the more necessary.

Forms of Technology Transfer

Technical control begins with the sender with the six key operations *S, O, R, T, G, D*.

The technology is transferred to the receiver; or in other words a control situation is created at the receiving point through the same six operations.

Two different types of transfer are evident. It can be a straight copy of the sender's objectives and equipment; trying to create at the receiver the nearest similarity to the sender's situation even to the extent of copying the sender's working methods. The same systems, jobs, recruiting standards, education, visual aids, working groups, and even planning. On the other hand, while trying to achieve the same performance as the sender, totally different means adapted to the receiver's culture can be used.

That takes us from the "heterograft," a grafting on to a different organism, to the "homograft," the cultivation and development of an integrated grafting. And analysis of these two types and their fields of application will be further developed later in this chapter under "The Actors in the Transfer."

Stages of a Transfer

The four main stages are

1. The percentage of national manpower
2. The performance level of the job
3. The degree of individual initiative in the job
4. The level of complexity of the technical system

The first two parts were discussed under "Working Groups" and "Satisfactory Transfer Conditions" earlier in the chapter, so we shall examine the third and fourth parts. Concerning the third we should recall a remark made earlier: the history of national and companies' industrial emancipation demonstrates an almost ineluctable sequence—manage, maintain, control, regulate, improve, install, invent. If a short cut exists, which permits a country to find another route through a technique that it has not mastered at even a simple level, then we know little about it.

It would be useful to mention here the recent example of new uranium enrichment methods developed by the West Germans and South Africans, neither of whom had previously built gaseous diffusion plants. But they do not seem to have the nuclear industry's sanction.

Without wanting to extend the debate, it is difficult to resist the temptation to cite Nietzsche on short cuts: "Apparent short cuts have always exposed humanity to great dangers. At the moment of learning that a short cut has been found, it abandons its proper route—and loses its way."

The fourth part leads to simplified technologies of which there is much talk, but little action.

The attractive idea here is to simplify the system to make it easier for the receiver to absorb. It may even be that in certain cases the sender will transfer obsolete technology to a developing country, which can be profitable in the receiving country because of less supply, cheaper transport, and lower salaries.

The idea, springing from the application of preventive ergonomics to the creation of the technical system, extends to the management system too; we have concluded many successful operations in this field.

It is better at the start to avoid complex management systems and use simpler solutions that will have a better chance of being accepted; it is also a more natural step towards more elaborate systems.

The Cultural Impact of a Transfer

There is a temptation to simplify the transfer by reducing it to a simple arithmetical operation: "You don't have this technology, we will teach it to you..." But this does not recognize the fact that between two individuals (one with the know-how, the other without), the difference is far more than a technical gap. The know-how gap spreads into relations with other people, their attitudes to problems, and even their professional and social values.

Take the example of a manager in a highly industrialized country. He is well informed in his speciality, be it law, finance, or engineering, but lacks any knowledge of data processing. Consciously or not, that lack reflects on his personality. And above all on his professional personality: he pleads for a simple pragmatism, remarks ironically, even condescendingly, on how narrow and esoteric data processing specialists are, or the enormous amounts of paper they produce, and how frequently it is useless. It is also reflected in his private personality: computers are responsible for mistakes at the bank and the slowness of social security payments, and he denounces them as being at the center of a threat to control his private life.

Now, take the example of a shepherd in a developing country seeking a job in a plant under construction. He lacks not only technical know-how, but also an understanding of the abstract, of how to interpret even the simplest technical drawings. His background is also at odds with an industrial society. He does not understand the importance of time, an essential consideration in all production systems. He does not realize why a work team should all be present at the same time for a meeting. The need to make machines work properly and alertness to breakdowns, are foreign to him.

The conclusion drawn from these two examples is that a transfer goes beyond relocating technology; it also encompasses the social behavior of the people involved, a part of their culture. By culture we mean "the complex whole that encompasses knowledge, beliefs, morality, law, customs, and all kinds of habits acquired by members of a society."

When a transfer takes place the culture of the people involved takes on a new dimension induced by the technology; it becomes part of an international culture which aids bilateral exchanges.

The Actors in the Transfer

According to their complexity, the technology transfer operations—from sender to receiver—can involve any number of actors, varying from two to many.

The simplest situation is where there are two actors, a group of receivers, well defined, but perhaps not yet in being, and a sending group, whose profile is more difficult to define.

The moment that these two groups go beyond a straightforward copy transfer, a third actor appears. He is responsible for devising an appropriate system for the transfer and to coordinate its implementation. This third actor is the technology transfer consultant.

When the transfer occurs simultaneously with the erection of a new plant, a fourth actor emerges—the engineer accompanied by builders and businessmen.

Before plunging into the various transfer situations (classified by the number of actors involved) it is wise to become familiar with the whole company of actors by looking at the cast. In order of their appearance on the scene they are as follows:

1. A company that seeks a certain level of technical competence, that is, the receiver of technical competence. The technology is implanted in a country and in a receiver's environment.
2. A company with a certain level of technical competence (the so-called reference technology) which wishes, with the help of subcontractors, to become a technology sender or lessor. The company possesses part of the technology involved: factories, structures, programs, studies, and so on. The technology is in the sender's environment.
3. A team of experts, creative and practical specialists in transfers, otherwise known as the technology transfer consultants (abbreviated to consultant). It can also be a team of transfer engineers (other terms employed are education or training specialists).
4. A company of technical engineers, capable of planning and realizing an industrial installation adapted to local conditions.
5. An equipment specialist capable of taking global responsibility for the industrial project.
6. One or more lessors.
7. Builders who belong to one and/or the other of the sending and receiving organizations.

8. One or several companies specializing in temporary labor and supplying work forces to fit the needs.
9. Training establishments coming from either sending or receiving organizations.

This cast takes no account of two situations which, although more apparent than real, might alter the number of actors. On the one hand, certain large industrial groups might contain several, or indeed all, of the transfer actors. Such legal simplification in no way alters their characteristics or competence. Yet, and probably more frequently, links are formed to reply to overall questions including, for example, the building of plants or the technology of how to run them (products in hand). In this case, it often happens that the operation is directed by a coördinator who directs the transfer, without having any actual competence in the matter. He must therefore bring in a company specializing in the necessary techniques and probably a transfer consultant too; and that brings us back to the distribution framework. Figure 3-3 illustrates the links between the technology transfer and the actors involved.

A Two-Actor Situation

One is the receiver, the other the sender. The receiver will usually be, for example, a company in a developing country already possessing plants (or in the process of acquiring them) seeking a certain level of expertise. It envisages achieving that with help of a foreign company. For the transfer to be effective, the foreign company must be at a higher level of technical competence than the receiving company. That favors technology senders in the same industrial sector and in the same business as the receiver. In other words, through the numerous functions it performs, it possesses the required experience. For example, if it is a question of setting up a naval construction industry, only foreign firms in that particular field have reached the level of technological mastery necessary to be a good technology sender.

It is not enough to say that the technology exists in the industry of the sending country. We must go further in analyzing the industry, being well aware of its specializations. A high-temperature furnace maker is not necessarily a qualified steel maker; a turbine manufacturer is often less skilled in getting the best performance from his machines than an electrical power producer.

Thus, two essential actors are needed:

1. A receiving company seeking a certain level of technical competence.
2. A well-placed sender company with definite technical superiority in the relevant domain. Called the "reference" company, it is prepared to sell its technical competence (assuming it exists in the company). The company

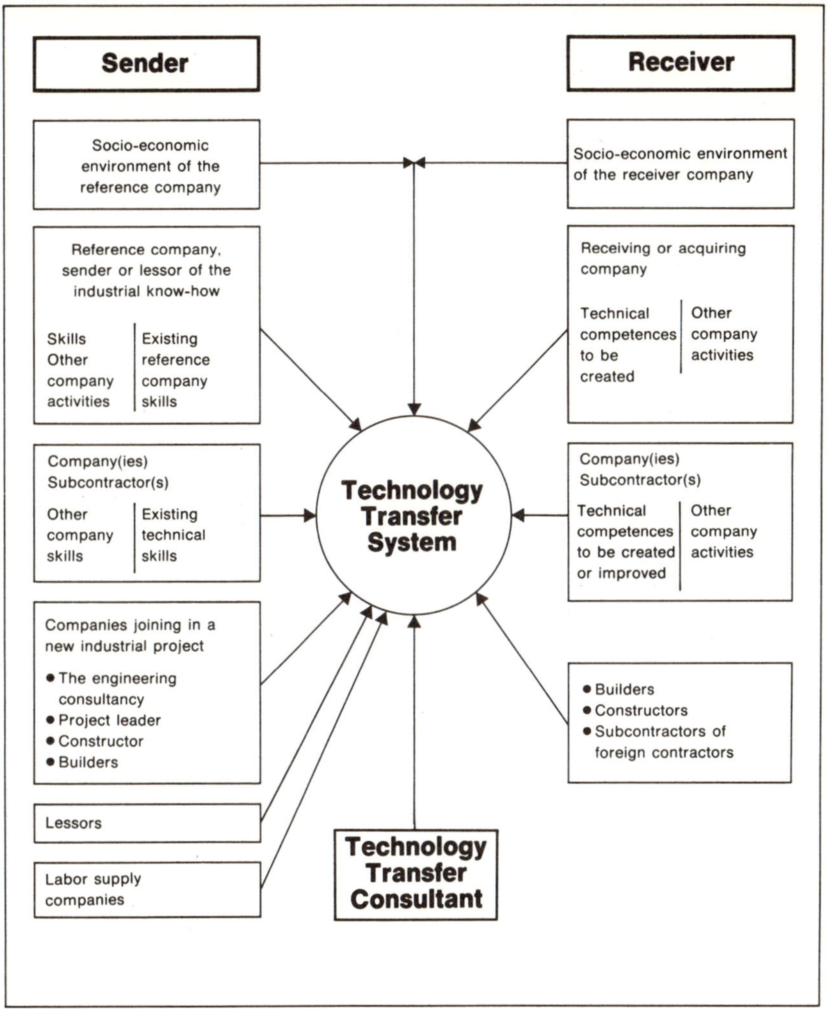

Figure 3-3. This diagram shows how the actors in a technology transfer are linked to the technology transfer system.

will often involve its subcontractors in the operation—the level of integration between manufacturing and maintenance generally being emphasized more in developing countries than in industrialized ones.

The following question deserves attention: Can a management hierarchy be established between the two roles of receiver and sender? The answer is twofold.

First, from experience we know that the transfer will not be successful unless there is a concerted effort between receiver and sender. Neither can do anything without the other. Both should be aware of their complementariness.

Second, if we were forced to draw a distinction between the two we should say that the sender has the major role. That judgment is based on one of the most brilliant transfer successes ever: Japan's.

In fact, only the receiver can properly

1. Choose the new technology by making the most of local aptitudes (after studies by various experts)
2. Orient his best people towards the new technology
3. Create around the new technology a competitive spirit, stimulating people to learn
4. Put into place the training system leading to and preparing for industrialization

Japan has cleverly taken advantage of these four opportunities and no one would deny their leading rank in industrial emancipation (which despite rapid growth still extends over 110 years). Saburo Okita explains this success in much the same terms as above: "In Japan the quality of the work force, the accumulation of capital, the social atmosphere and other basic conditions, all favored the effective absorption of foreign technology."[4]

Christian Sautter recalls the strategy Japan created at the turn of the century to catch up with burgeoning industrialization.[5] It concentrated on light industry, particularly silk, which between 1870 and 1929 accounted for between a third and one quarter of Japan's exports. But why choose silk as the sole product on a national scale? Because the industry, based on the breeding of silk worms and the emptying of the cocoons, and the domicile-based Japanese cottage industry, was perfectly adaptable to the first two industrial stages. Nonetheless, some investment, though not too high, was needed. The installation of modern workshops was realized with the help of French and Italian specialists who were the source of the technology transfer and the industrial culture.

It is worth noting that, while playing the light industry card and discovering what modern economists today call "the Hong Kong model," the Japanese were astute enough to lay the foundation for a still far off, but ambitious, future. Their turn-of-the-century embryo grew into today's heavy industry. Japan had thus implanted the seed of a new culture—an advantageous jumping-off point for later expansion.

One last comment should confirm the preeminence of the technology receiver. There are examples of successful transfers where the receiver is very

---

[4] *Ibid*, p. 12.
[5] Christian Sautter, *Japon, le Prix de Puissance*, le Seuil, 1973, pp. 17, 18.

active and the sender passive or even hostile. The most notorious example is industrial espionage (a kind of deviant transfer which we can hardly recommend, but which exists all the same). Certain successes are well-known. On the other hand, there are no examples of a successful transfer with a completely passive receiver.

As Algerian economic journalist, Mohamed Brahimi stresses: "The technology must not be received passively by the Third World. It must be sought."[6]

But returning to the relationship between the receiver and sender, between seeker and seller, how does the transfer operate? There is a great temptation to bring together in the reference company all the apparent elements leading to technical competence, and to graft it all on, like a photocopy, to the receiving company.

Among these elements are some that have already been the object of commercial transactions. These are patents covering products or manufacturing processes. Others, just as vital, but less tangible, were revealed earlier:

S —The technical system (man-machine or management) revealing the tasks to be performed
O—Organization of the macro- and microstructure; organizing the tasks into jobs that will suit each worker
R —Recruitment revealing the best equipped for the tasks
T —Training of each worker up to the level required by the job
G—Group training showing each member how to go beyond his individual role
D—The dynamism of the whole

This simple solution can give useful results as long as the difference between the companies is not too great. And it should be emphasized that numerous elements to be transferred are not explicitly revealed in a transferable form, but are often customs or traditions contained in the memories of various top executives.

Lowell Steel of General Electric stressed that point in an address to an EEC seminar on technology transfer, held in Geneva in July 1975: "The information (to be transmitted to the receiver) is dispersed throughout the sender's structure, based on his internal needs. It is frequently mixed up with other information which is not part of the transaction and one often finds that relevant documentation simply does not exist. The actual practices followed in the workshops rarely conform entirely with the written specifications of plans."

The circumstances of a direct, carbon-copy graft are laid out in Figure 3-4. At the center of this diagram is the sender system and at the top the receiver system,

---

[6]Mohamed Brahimi, "Le transfert de technologie," *El Moudjahid Economique*, February 3, 1976.

"A" in the two-actor transfer. Whether receiver or sender, each system has three levels: a socio-economic environment, a company, and finally, that part of the company concerned with the technology to be transmitted. The most important characteristics for each level are specified in Figure 3-4. The differences between the two first levels, between receiver and sender, may not be too pronounced because identical transfers are possible, as earlier stated. To use a simple analogy, we can say that the two companies are on neighboring wavelengths and thus the transfer is mutually satisfying.

The most common case can now be confronted. That is where the differences between the socio-economic environment of the receiver and the sender go beyond a certain level, and may only be intuitively perceived.

Experience shows that a simple photocopy transfer sometimes does not graft on and can lead to negative reactions, such as trainees' dissatisfaction with their training, failures in the examinations, and a rapidly alarming absence of technical competence. Other problems can arise also, such as unforeseen incidents, an absence of spare parts, growing number of rejects, and long down-time.

But here we are no longer strictly speaking about a transfer. That is not to say that the work put in is inefficient, it is simply of a different nature and, under those circumstances, it may be better to speak of technical aid: managers and training specialists from the sender company can play the same role in the receiver company as well as they do in their own company. They will thus ensure, as long as they are there, that a part of the receiver company functions properly. But that is all they can ensure, for the cultural differences between sender and receiver prevent an effective transfer. When technical aid is withdrawn, the level of competence drops.

Unfortunately, such circumstances are frequent and are accompanied by disputed legal claims which prevent a dispassionate analysis and hinder a correct understanding of the causes of the failure. Several explanations can be put forward for that: the sender might blame the workers' resistance to change, or the lack of a deeply enough ingrained industrial culture. The receiver might regret the insufficient commitment by the sender, but to avoid future hazards will insist on more and more restrictive clauses in the contract.

And yet the explanation of such failures seems simple enough. The six previously defined characteristics of a technical mastery situation are largely influenced by three factors contained in every company (see Figures 3-4 and 3-5):

1. The socio-cultural environment
2. The structure of the whole enterprise including the management system on which it is based and the people who run it
3. The degree of penetration of the technique involved

The six elements, technical system (*S*), organization of the macro- and microstructures (*O*), recruitment (*R*), training (*T*), and group training (*G*), and

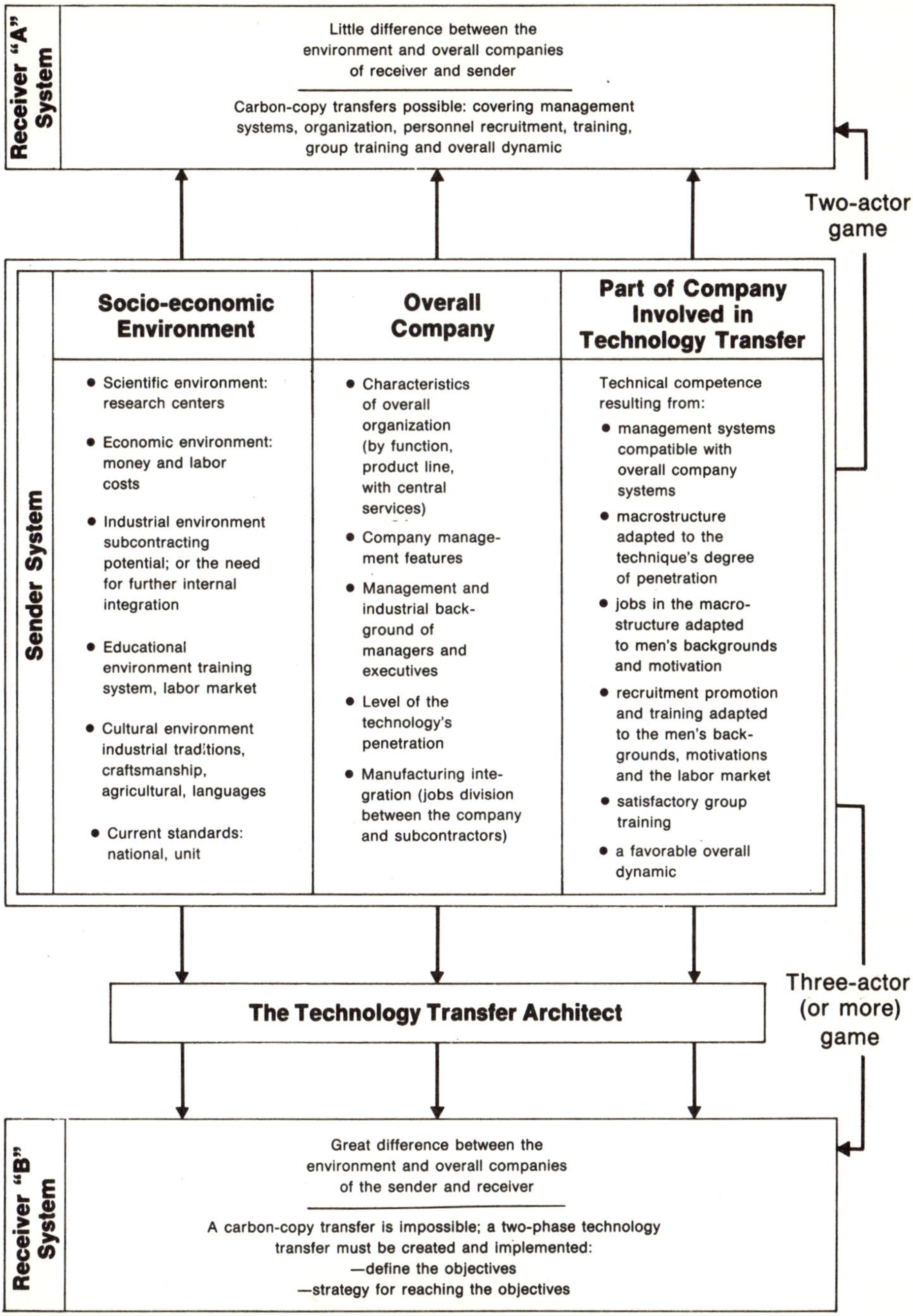

Figure 3-4. Shown here are the circumstances permitting or prohibiting a carbon-copy technology transfer.

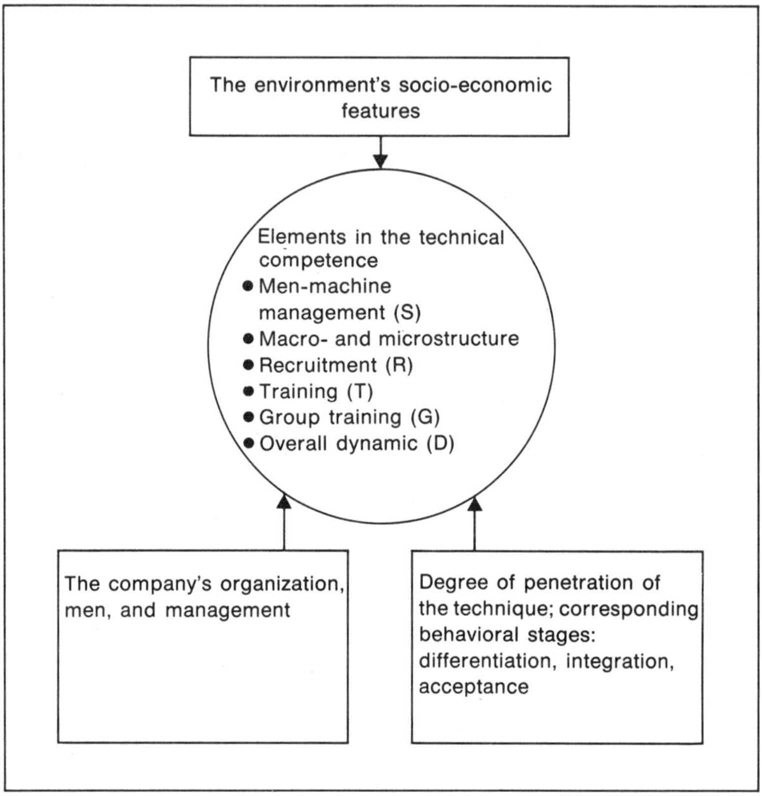

Figure 3-5. Influences on a company's acquisition of technical competence.

overall dynamics (*D*), do not constitute overall a universal optimum; otherwise known as Taylor's "one best way." The transposition is sometimes possible, but only within a certain minimum level of differences.

Beyond that it is no longer a question of the complete copy. On the contrary, the technique for mastery by the receiving company must be expressed in terms adapted to its organization and socio-economic environment. A real transfer system, reflecting the originality of the situation grouping the six characters S, O, R, T, G, D, and adapted to the receiver's needs, must be thought out and implemented (see Figure 3-4, receiver system "B," three or more actors). And to borrow an expression from a report published by the Dag Hammarskjoeld Foundation,[7] "innovative and nonmimetic solutions" must be sought.

---

[7] Report presented at the extraordinary assembly of the United Nations from January 1 to September 12, 1975. Quoted by Pierre Garrigues, "A Technology at the Service of the Future Single World," *Synthesis* 11, 1976.

We should like to mention a personal experience concerning the links between the educational programs and the basic culture of the men, including their life-styles. An identical copy training program leading to the professional aptitude certificate for lathe and milling operators was set up in October 1960 near the works of the Société des Mines de Fer de Mauritanie (Miferma; see Appendix 4). It was similar to French three-year courses. Production start-up was planned for 1962. The previous year, March 1961, we had visited the training center.

There were numerous gaps in the system: the trainees arrived two years too late, they were too few in number, and no preparations had been made for the most urgently needed of all jobs, those needed to keep the plant running. The course had started three months before our visit with only twelve apprentices, whose recruitment was extremely difficult because of the difficult entrance examination. Now only three remained and the training officer drew disillusioned conclusions about the impossibility of a Mauritanian to adapt to an industrial environment.

Yet one year later 800 adult Mauritanians trained by us took up 70 different kinds of jobs in the company. What had happened? We made no attempt at an in-depth training so different from the life-styles of nomads who were to work there.

We had carefully gathered together all the assets in their culture and customary environment and had used those assets in each of the different training stages.

The problem was thus not insoluble: simply the wrong solution had been applied.

A Three-Actor Situation

That shows up the need for a third actor who can conceive and put into effect an "original" transfer technology system adapted to each situation. Figure 3-6 details the skills this third person, the technology transfer specialist, must possess.

The specialist is rarely an individual on his own, as the required skills demand a structured team. Depending on the case he can be independent of both sender and receiver, attached to one or either of them, linked to a project coordinator, or to a whole-project engineer. What is essential is that his skills, which differ strikingly from the operational executives or technical engineers, must be based on serious practical experience. The importance and independence of his role must be acknowledged.

The intervention of the three actors unfolds as follows: the consultant, previously informed of the receiver's characteristics through a situation study, gathers, in cooperation with the sender, everything considered the raw material of the message to be transmitted. He will quickly assemble a mixed sender-receiver core of executives with whom he will create a technology transfer

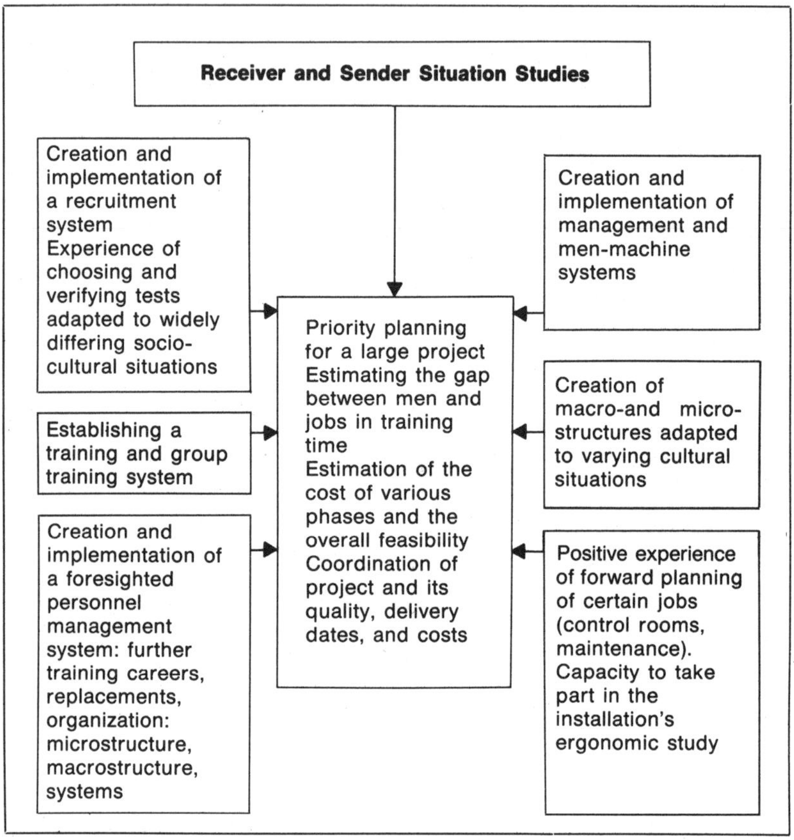

Figure 3-6. A summary of the main skills of a technology transfer consultant.

system that foresees the various transfer steps and analyzes the jobs to be done: management of men and machines, micro- and macrostructure, and the elements involving the men—training, and internal or external recruitment.

The interplay of these four elements leads to harmonious men and job relationships. Two final transfer elements complete the global system. First, group training where each worker can rise beyond his job through appreciating the role of his neighbor and the interactions with his own job. Secondly, by foreseeing the growth dynamism, particularly where workers' progress is concerned, which must coincide with training steps and improvements in organization.

The sender comes in, with his technical strengths, in various forms. He may admit, either into his plants or training centers, individuals specially chosen by the receiver. They should preferably have a precise information goal in mind, for example picking up the various elements in a training course. The sender may send some professionals to the receiving company who will work in close liaison with the consultant devoting their time to management tasks. For example, running the installation in the interim before it is taken over by the nationals, or the preparation and delivering of training sessions.

The difference between the role of the seller of technical mastery and the technology transfer consultant (who in fact is the adapter between sender and receiver) must be underlined.

The seller puts the various elements of his industrial competence at the disposition of the operation. They will be such things as documents, organizations (information centers), and trained workers. It is not the seller's job to worry too much about the details of how these elements are to be adapted to the receiver's characteristics. But it is the central preoccupation of the consultant. He takes several well chosen elements from the receiver as components of the technology transfer system and completes them with other new elements.

This comparison of the roles leads us naturally to a question already raised by the acquirer-seller couplet: can one set up a staff hierarchy among the acquirer-seller-consultant trio? Or rather, as the role of the acquirer is not being challenged, who, between seller and consultant, exercises the dominant role?

In fact, it all depends on the situation. The importance of the consultant's role grows according to the socio-economic gap between the sender and receiver. The role of the lessor grows with the complexity of the technology to be transferred. Above a certain socio-cultural difference level, the consultant's importance preponderates. It is his job to decide whether to use existing programs or to create new ones closely adapted to the native population.

It should be noted that whenever the technology being transferred is a management technique, usually well known to the consultant, the schema can be simplified; the consultant can then marry his role with the seller's. We thus arrive back at a two-actor situation: the acquiring company and the consultant.

Binational Industrial Cooperation

In the two- or three-actor set-up a favorable transfer situation can be perceived in the following circumstances.

It is clear that any transfer demands recognition of the characteristics of the sender and the receiver, which is in no little way dependent on the extent of the technologies to be transferred. However, it frequently happens that a company, particularly during its creation in a developing country, needs transfers stretching over a wide range of production and management technologies. The transfer will, therefore, be much simpler if it takes place between the same sender and receiver, using the same preliminary studies, and taking account of the growing cultural closeness between them. Such a circumstance arises where a cooperation agreement (which may take many legal forms) exists between a company in a developing country and a similar company from an industrialized country.

The agreement could cover such things as market studies, product ranges and sales prices, manufacturing capacity, placement and types of plants, orders for constructors or engineers, organization of the sales network, financial management, and so on. Sometimes it can lead to a division of production facilities between the two partners, resulting in the relocation of some of the sender's workshops in the developing country.

Certainly such cooperation demands a forward-looking attitude from the partners that may not always be easy to find. Each partner may harbor certain fears about the other, even if the sender is correctly and adequately paid and the receiver is capable of using the technology independently.

Nonetheless, binational agreements, aided by technology transfer consultants where necessary, are a realistic and rewarding means of acquiring technical aid and the effective transfer of the range of technologies necessary for all growing industries.[8]

A Four-Actor Situation

This situation arises when the transfer accompanies a new industrial project. The link between technology transfers and the building of new plant is so close that many people would be surprised to see the technical engineering company being placed only fourth behind the receiver, the lessor of technical mastery, and the transfer engineer (a phrase used in preference to consultant to distinguish their complementary roles).

We should first recognize that the actors are defined according to the actual placing of their specialization rather than in any legal ranking. A technical engineering company with a technology transfer department would obviously figure second in the three-actor situation, with a double reason for being in the present situation.

It happens more often that one would think that the transfer is accomplished with only the three actors listed above. That will be the case particularly where there is no new industrial investment accompanying the transfer such as with

1. The starting up of a training system within the company or at a professional level.
2. Improvement in the technical competence of the crews working the existing plant.
3. Replacement by nationals of contract-expired foreigners.
4. A management function transfer, where machines are rarely, if ever, found; eg. strategic planning, investment decisions, personnel manage-

[8] The following formula sums up the scope of the industrial cooperation:
$IC = K_1\,AT + K_2\,TT$
with IC = industrial cooperation, AT = technical aid
TT = technology transfer, $K_1 + K_2$ between 0 and 1

ment, marketing, and supply. (We have even seen examples of this where the transfer was carried out by two actors—the receiver and the transfer engineer.)

Yet if we take the case where the transfer accompanies new industrial investment, it is clear that the engineering companies are essentially equipped to supply the market's traditional needs; that is, creation and start-up of plant fitting the entrepreneur's needs and demands.

But while technical engineering dominates the material aspects, in all their forms, human problems demand the skills which are the speciality of the technology transfer engineer. Nonetheless, if it needs to be repeated, the role of technical engineers in a transfer is far from negligible.

As with transfer engineers, the technical engineer is faced with different demands with each new project. On the one hand he has experience of similar projects through having been involved with them in developing countries. He will also possess reports of model solutions, which he may be tempted to apply identically. Yet, on the other hand, he sees many differences between the new plant's environment and the original "reference" environment. He will, therefore, recognize the need for a deeper knowledge of the geographic, climatic, economic, and social factors that beg for a new treatment vastly different from the model solution.

So engineering, both technical and transfer, seems to be a harmonizing factor between industrial plants in developed countries and the socio-economic and geographic realities of developing countries.

Technical engineering's influence on the project centers around raw materials, products, processes, creation, and implantation of the plant. Technology transfer engineering influences the operational structure: jobs, selection criteria, group and individual training. Cooperation between the two kinds of engineering can prove fruitful in at least three cases.

First, it is clear that the transfer will be simpler and more effective if the plant's technical basis complies with the technical realities of the country. For example, potential trainees will be more highly motivated if the plant's products correspond to market needs, or if it injects greater added value into raw materials without sending them abroad.

On the other hand, a strong bridge must be built between the way the initial studies and the actual transfer are carried out. The latter demands information about the plant starting with an overall picture, then becoming more detailed at each stage of its development. The various steps to start-up must also be precisely understood.

There also exists what is often called a no-man's-land outside the installing of the plant and training of workers, where a rewarding cooperation between the technical engineer and the transfer consultant can take place—and that is creative, or preventive, ergonomics.

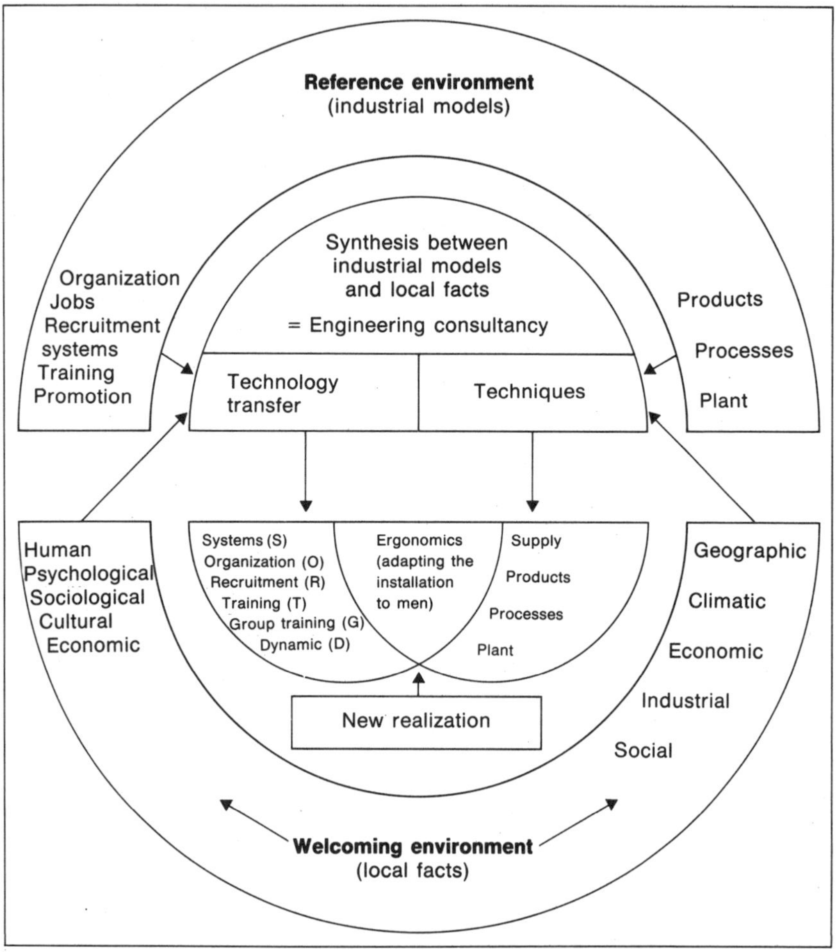

Figure 3-7. Complementary roles of the engineering consultancy and the technology transfer consultancy.

Ergonomics is involved (too often as a corrective) in adapting jobs to the workers. That adaptation can be achieved by changing the abstract elements such as structures. But it can also influence such tangibles as the plant and its installation. Ergonomics, therefore, occupies a privileged place during the drawing-up of a project, making one aware not only of standardizations which can simplify work for the entrepreneur (valves, contact points, motors, even very large machines), but equally, of the many pieces of operational equipment and the adaptation to a human scale of a remotely controlled plant. We have even seen that ergonomics can extend to the whole technical system and to the adoption of simple, or soft, technology.

According to each case (that is according to the competence of each man) ergonomics can be applied by the technical engineering or by the transfer engineering, or better still, by both of them acting together.

A More-Than-Four-Actor Situation

Going beyond the four actors, and still within the scenario of a new industrial construction, other organizations in a by no means negligible role, will also be present:

1. The lessor, sometimes different from the reference company.
2. The overall coordinator who covers all technical engineering.
3. Equipment builders and entrepreneurs: as with technical engineering, these actors must give advance notification to the transfer engineer (through their schedules), of equipment, its installation, running and maintenance systems, and dates compatible with an ordered progression of the transfer operation. Moreover, the constructor, must be capable of receiving certain trainees, putting training tools at their disposition, and accepting certain receiver company crews to work among its own assembly crews.
4. Perhaps a temporary recruitment bureau, and/or an ad hoc management and recruitment organization, supplying and managing a certain number of foreign-recruited specialists, if they do not already exist in the receiver company and cannot be spared from the sender company.
5. Several training establishments, preferably from the receiver country.

As soon as the project grows to a certain size, such as described in Figure 3-7, it is often preferable to create a special structure, in the form of an operation transfer group, bringing together the various actors concerned with the success of the operation.

4

A Proposed Method: The Eurequip "T" Method

Is there a method that can be applied to the various problems posed by technology transfer, while respecting its diversity? At the beginning of our work we never imagined that it would be possible to cover the many different avenues by one overall formula.

It is with some hesitation, and lacking of a more precise term, that we envelope the range of our services in the word "method." Yet in admitting that there can be a method, it is expressed in a logical ten-point sequence.

At the method's center—that is, between points 3 and 8—we find the six operational keys to a transfer: *S* for the system, *O* for organization, *R* for recruitment, *T* for training, *G* for group training, and *D* for the overall dynamic.

But first comes point 1, the indispensible preliminary study which splits into two parts: the receiver and at the sender. Point 2 is the constitution of the core of the transfer, that is the selection and training of the project leaders.

At the other end, point 9 is devoted to laying down a means of evaluating the success of the various transfer operations and to modify those operations if they do not meet target objectives.

Point 10 is concerned with the transfer strategy and its expression through an overall plan.

Preliminary Studies

The Receiver's Situation Study

When a steel-making plant in an agricultural region shut down, local authorities prepared the way for new industries on condition that priority be given to the newly-made jobless. A fertilizer plant was planned. What are the employment characteristics of the steel plant personnel? Which employees are most adaptable to the new fertilizer plant? An adult training organization prepares retraining classes. Which classes are most suited to the workers and the new jobs? What further training will be necessary? Which employees can be taken on from the steel plant; which employees must be recruited elsewhere, and how? And finally, will the industrial changes taking place create any complementary functions between the planned fertilizer plant, the existing regional industrial web, and other industries to be set up there.

> *Technology transfer starts in the receiver country region, or company with a situation study; it must take into account the available population and existing training systems, as well as the industrial structure.*

The aim of both the receiver and sender studies is to formulate the problem by setting out the quantitative and qualitative objectives, and through analyzing the constraints, to identify the actors in the transfer and to rough out the major lines of the solution. Those rough lines will be smoothed out in following steps starting with an overall general estimate of what is required by reference to a similar project in another country.

We usually start with an examination of the work force insofar as they satisfy certain policy decisions (concerning nationality, age, nearness to the plant, freedom to move from a neighboring factory) and aptitude for the jobs foreseen (level of education, experience, and so on). At this stage the gap between the worker and the job is expressed in time and possibilities for training him. A nomadic population as we have already seen, would hardly stick with training that lasted several years. Sometimes the studies on labor sources can extend to neighboring countries with whom agreements for labor and even management, might exist.

This first study should provide

1. A rough idea of the labor source, that is, for such a job we can find such and such labor.
2. A rough guide to the foreign labor sources, whether from neighboring countries or from mature industrial states further afield. (The possibility of acquiring part of the labor needs through one or more of the "reference" companies will be examined later.)
3. An outline of the composition by nationality of the future labor force.
4. A preliminary indication of the characterisitics of each source: knowledge, technical experience, cultural levels, interests, language, possible conflict with other racial groups, and so on.
5. A preliminary study of selection and recruitment adapted to each source, and probably some initial thoughts on the selection methods available and their validation.

The teaching system, whether technical or general, must also be made clear, including the training centers in the receiver company. This permits

1. A better knowledge of the labor sources from certain teaching establishments.
2. The communication of educational experiences.
3. And finally the location of certain training centers that can help solve parts of the overall technology transfer problems.

If the receiver company already exists and the new technology can simply be grafted on, the workers, company organization, and skills studies must nonetheless be carried through in detail. In particular we shall examine the best way of inserting the new technology or plant into the organization (see Table 3-2).

Finally, the receiver's situation study must also cover the work environment to ensure that it is suitable to the labor force. Whether it is a question of accommodation, direct or indirect wages, even force, all these factors, which may seem subsidiary, are in fact vital for labor force stability and for the transfer's success.

The Sender Study

A national company in an African state awarded to a European contractor the construction of a refinery complete with a recently operational oil well. Shortly after the order for the turnkey plant, the problem of the labor force and the technology transfer arose. A study of the receiver's situation was successfully carried through, but was not sufficient. The sender's environment must also be studied in its various forms, including:

1. Project leaders and equipment manufacturers who are responsible to the plant buyer for the technical specifications, construction and operational programs description and plans for running and maintenance, and for knowing where similar plants are already in operation.

2. The reference company, which preferably should possess a similar plant that can serve as a source of technical mastery for the new refinery. Beyond any contractual problems, this presupposes the acceptance of African trainees in its training centers and plants, as well as the availability of managers and training officers who can be detached for several years.

The situation study must cover the reference environment; that is the potential sender, any similar plant, the technical engineering company, equipment manufacturers, schools at every level, and the market for the technical training of workers.

The study must go beyond a simple list of the sender's technical competence in order to detail potential know-how sources, the teaching methods to be used, and those who are ready and willing to undergo training and its disciplines.

As we have seen the sender can be a diffuse entity spread throughout a whole country. The study must, therefore, concentrate on similar kinds of plants operated by a reference company or certain subcontractors, the construction engineering company (in certain cases the project leader), equipment builders, technical schools that can be linked to some of the transfer actors or their environment, and finally the labor market. Special attention must be paid to future expatriate employees and their chances of finding work at the end of their contracts.

Similar plant (or the reference company). These are the technology sources and finding them becomes harder the older the company is because the technology is dispersed among a large number rather than being concentrated in the hands of one person. One should also look at the possibility of personnel detachment: the use that can be made of its specialists and the attitude of the company and its staff to this possibility. Finally, the possibilities for receiving trainees; their experience in this area merits attention.

The engineering company. This is the company that usually prepares the market and drafts the contracts. Yet, while they are often prepared too quickly, these clauses nonetheless list the key points in the transfer itinerary. They can be such details as the placing of equipment in the control room, starting-up instruction for a lathe, routine inspection procedures for machinery, or such information as is only available after plant start-up. In fact, they are necessary nine to twelve months before then to be included in the personnel training. (Appendix 2 provides thoughts on the setting up of a factory and some advice on the ergonomics of creating work conditions on the drawing board.)

The equipment manufacturers. They will willingly comply with requests for information in good time, if they are aware that the future plant operator aims to put as much careful preparation into his labor force as the constructors put into the manufacture of equipment.

Technical schools. These are usually a complex group extending from factory apprenticeship schools to schools of differing technological levels, including professional training centers. Some of the schools can accept trainees. The majority possess very useful education material. But rarely does one find a school so closely adapted to the specific problems posed by the transfer that it can be grafted directly on to the receiver.

The labor market. This labor source will be necessary if, as frequently happens, the reference company, the engineering company, and the constructors do not possess sufficient personnel of the right quality to supply technical assistants. This market in situation and evolution, must be well explored, including the possibilities of widening the search for labor in other industrialized countries.

Engaging a personnel selection company, and perhaps later a personnel management specialist is recommended. The other transfer actors are often competent at selection (the consultants for example) but rarely have the skills for managing expatriots working on fixing duration contracts.

A provisional technical assistance cost will be estimated at this stage as well as the marginal cost of replacing the lowest grade technical assistant with a trained native worker.

Selection, Training and Role of Technology Transfer Leaders

Immediately after the Second World War French industry became aware of the delays that had accumulated since 1930 and particularly since 1940. Production missions to the USA were organized, with government approval, to amass American industrial performance data. From these missions emerged leaders who drafted reports, organized think sessions and spread the message of what had been learned after filtering out the elements in American culture that could not be transplanted in France.

A large part of the American experience was, therefore, broken down, transposed and assimilated thanks to the efforts of a large number of technology transfer leaders. Their influence extended among widely differing industries and to all parts of the country. The French "miracle" of the 1960s is partly explained by the grafting of American experience onto a rich, unrealized, human potential.

The transfer relies heavily on the existence of training executives to link sender and receiver. The sender executives must learn the technical, cultural, and organizational situation of the receiver and the receiver executives must learn the facts of the industrial techniques to be transmitted.

Technology transfer is, above all, a matter of communication. And, if as frequently happens, the cultural gaps between receiver and sender yawn too wide, there is little alternative but to bring in "interpreters" to identify the message in the first environment, then transmit it to the second in a comprehensible, assimilable way.

Curiously enough, it is only recently that we have become aware of the difficulties of communication between people. That difficulty is well-analyzed by Pierre Jouven:[9] "When I was fortunate enough to work in Africa 20 years ago I soon realized that the thinking processes of Africans were totally different from ours, and that it was impossible to find the solutions to some problems simply through discussion. We were already alerted to this problem by Frenchmen who had lived in Africa for some years. But some years later when I came into contact with Americans, I saw that there too, we sometimes came up against a frame of mind which was incomprehensible to us. And yet we were dealing with long-experienced industrialists, much closer to us than Africans from a completely different culture. Similarly today, some of us, arriving at a certain age, find ourselves completely cut off from young people whose mental processes are totally opposed to ours."[10]

Where do the indispensable technology transfer "interpreters" come from? Ideally they are found in the receiver's environment but that is not always possible. Another solution is to create mixed groups drawn from both receiver and sender. (Appendix 3 provides several examples of technical training among workers—many of them illiterate—that succeeded thanks to close collaboration between European instructors and assistant instructors from the trainees' environment.)

The best preparation for their jobs that we can give to these leaders is to give them responsibility (after proper training) for the preparation and implementation of a technology transfer system. It should cover the organizational aspects—*S* and *O* operations—as well as training aspects—I and G operations. (Appendixes 3 and 4 are examples of positive participation by future instructors in the creation and preparation of the training program and course classes.) Their major tasks will be the collection of basic information from the reference companies and, most important, compiling the training sessions which will be adapted to the educational level of the future trainees.

The drafting of the collected information into training programs will always benefit from a mixed group. If necessary it can be left to leaders from the reference environment as long as it is controlled by a group from the receiver environment. Naturally, after preparation, the receiver leaders are given responsibility for applying the program.

It is advantageous (though rarely simple) to go even further and furnish the receiver leaders the management of the group that has been formed.

That latter course is difficult where a training center is concerned, but quite normal when preparing the labor force for a plant. Thus, the transfer leaders are found in local management, which is responsible also for training; and they can follow it up by further training after plant start-up. (Appendix 1 cites the example of common to plants under construction and of training leaders.)

---

9 Former chairman of the multinational corporation Pechiney—Ugine—Kuhlmann, whose American branch is Howmet Aluminum Corporation.

10 Extract from an article "La communication, clé des problèmes de l'entreprise," *L'économie*, n° 1217, June 9, 1975.

Creation of the System

From the moment that the preliminary studies for an advanced technology study center destined for a developing country begins, consideration must be given to the standardization of scientific and technical equipment.

Standardization applies on the one hand to laboratory and workshop equipment; on the other hand to subsidiary construction equipment such as lighting, switches, doors, curb stones, taps, and pipes.

Such an analysis can reduce the number of supply parts by a ratio of 3 to 1, and in turn that simplifies

1. Construction and assembly work by local companies
2. Use and maintenance of all kinds of equipment
3. Supply and stock management
4. The overall transfer of a range of construction research and maintenance technologies necessary to the proper working of the enterprise

Yet, when aiming for a certain objective, it is rare that production and management can adopt certain solutions without changing them in some way.

Choices can be made at the overall macromanagement stage and at the detailed micromanagement stage. These choices will influence the number and difficulty of the tasks to be allocated to the employees and, consequently, the transfer of technology to them.

The system in all its complexity, whatever its nature, is rarely completely imposed at the beginning of a project. What is usually defined at the outset, or more exactly during the initial study that sweeps together the various project actors, are the performance levels and the objectives that can be realistically reached as a result. They can be for example:

1. Production of a certain number of objects within a production system
2. Machine down-time in a maintenance system
3. Familiarization, forecasting, and improvement of the cash-flow for a company financial system
4. Mastery of a complex problem such as traffic control by traffic lights

Yet, in the same way that we can talk about the macro- and microstructure through the organization (O) and the macro- and microteaching in training (T), so we can refer to the system's macro- and microcreation. (In an urban context the equivalent disciplines would be town planning and architecture). It is interesting to examine the degrees of freedom of action at the two levels, whether it be man-machines systems, or management.

Man-Machine Systems

The term can refer to laboratories, sometimes to work-yards, and frequently to production units.

In that context the macroconcept concerns the choice of processes, production schedules, heavy machinery, setting up of the plant, and even the extent of mechanization and automation. The various choices are detailed in Appendix 2. They are the heart of industrial engineering's role and thus situated at the frontier of technology transfer.

So, we will restrict ourselves for the moment to mentioning various research paths, such as

1. The transfer to developing countries of processes inappropriate to industrialized countries for reasons of high raw materials, energy transport, and labor costs. It can even imply transferring a plant from an industrialized country to a developing country.
2. Simplification of productive machinery because of different economic standards. For example, a steam boiler or turbine will have a much simpler system for recuperating heat at low pressures near a natural gas field than in an energy-importing country.
3. Orientation towards simplified technology, otherwise known as soft or intermediate technology—notably that proposed by Professor E.F. Schumacher, founder and head of ITOG (Intermediate Technology Development Group).

Jobs and roles can be significantly simplified by research in three areas.

Standardization. During a visit to a large overseas plant we noted 22 identical boilers and 18 identical turbines. That fortunate situation, which made maintenance and spare-parts stocks so much easier, was not simply good luck. It was the result of the engineering choice; rather than looking for partial optimization of each machine, it reached instead for an overall optimum use including running, maintenance, and stocks.

Machine installation. The complexity and danger involved in jobs begins to grow at the drawing board stage without the project leader always being aware of the effects of his simple decisions on the daily work of men over a period of years; and that applies whether the jobs are operational, maintenance, control, or even security.

Installation of measurement and control instruments. Operating one, or a group of machines can be directed from a point far from them in the form of a control room. If it is true that the layout of instruments and switches on a car dashboard can affect comfort and road safety, then it is even truer for control rooms housing hundreds of instruments. The correct grading of equipment, grouping of dials and corresponding command switches, and preparation for quick, sure intervention in abnormal circumstances, all facilitate running procedures.

We have already remarked on the usefulness of large mock-ups for training operators; for example, flaws are sometimes revealed when certain conditions are introduced in the mock-up. These often result from the fact that the control room was conceived in a static, instead of dynamic, way as a part of a man-machine system with a variable running order.

Management

The macroconcept involves the system's level of complexity. As the complexity level rises, so a growing number of variables are created that present forbidding barriers for the noninitiated.

The management set-up can be both the best and worst of all possible worlds. The best if it fits in well with the culture of the transfer receivers and the company, because then it enriches and consolidates the culture, permitting good individual and collective performance. But the worst if the gap between the system's ambitions and the men's skills is too wide; in that case over-complexity leads to rejection. Rejection can sometimes be camouflaged by the appearance of a properly working system, when it is not in fact really being used by the normal operators.

So agreement from the beginning as to the level of management and the company's culture is highly desirable. Naturally, that agreement does not exclude a more complex management system as the cultural level of the company rises. A dialectic concerning the progress of the system and the men is developed in Appendix 7.

The management microconcept also deserves attention. A management system can be just as esoteric in its entirety as in component structure. That can apply as much to an explanatory table as to a questionnaire. But it can also be applied to the scale of the undertakings scrutinized and the way they are presented.

An example is the introduction of management control in a plant. It often fails because economic standards are imposed on production and maintenance staff. To them a monetary unit is a poor standard for judging the physical aspects of the work. Management control is much more easily introduced if we take examples from the backgrounds of the people involved. In that case the amount of waste, production levels, breakdowns, and capacity, are naturally accepted as rewarding goals to aim for.

The way the standards are presented plays an important part in their successful adoption: thus, graph curves speak more eloquently to the engineer and the foreman than an esoteric list of figures.

Organization

As we have already noted, organization is the logical follow-up to the creation of the system or systems. Throughout the technical system arise jobs that must be allocated to workers. It is, therefore, the human element that guides the organizer in setting up the microstructure, or, in other words, the number of tasks will define roles and jobs. However, there is a danger that the microstructure may be attached to an anarchistic overall structure which grew up, rather like some suburbs, from unplanned, individual developments. The microstructure must therefore be written into a macrostructure that fulfills the major tasks essential to the system.

Adaptation of the Macrostructure

A French company wants to introduce strategic planning into its management thinking. The board defines certain essentials: avoid a planner's ivory tower and try instead to involve a growing number of managers into the practice of strategic management. The exercise demands two principal actors—the company and the consultant. And a major question arises,—should the overall organizational structure be modified, and how?

Several solutions are possible and are found in different kinds of companies. At one extreme is the well-defined central planning department, but with the accompanying danger that it will come to see planning as its inviolable preserve. At the other extreme is the maintenance of the *status quo* without a specialized department. The danger there is a threat of spreading the project over too wide an area and losing its way.

Finally, an intermediate solution is adopted. An entity is created to lead working groups, control their work, communicate necessary data, and aim each year for progress in the quality of the plan and the number of people involved in its establishment. Being essentially an operational organization the body will one day disappear when the strategic methods are solidly anchored in the company.

> *Rarely will the technique be so inflexible as to permit only one structural form. Several solutions are often conceivable. The choice of a structure adapted to the actual culture of the group involved, and its possible extension, is a simplifying factor in the transfer.*

Two concrete examples reveal the scale of such choices.

One is the place of electricity in a plant. In heavily industrialized countries different organizational structures will exist depending on the industry.

In certain cases everything involving electricity in the plant (production, transport, distribution, electrical machinery) comes under the specialized service department, responsible for operation, maintenance, and control. In others, it is lumped together with fluids and products, and is therefore divided between three service departments built around one function (operation, maintenance, control) which must be familiar with the whole plant.

Plant organization usually evolves from differentiation to integration.

The first takes into account electricity's specific nature and danger; it is contemporaneous with the creation of specialized schools for electricians.

But the second cannot be adopted until the company has employees at every level who can deal with electricity among other duties. This latter situation offers numerous operational, maintenance, and career advantages to electricians. Yet it cannot be applied until the employees' technical skills have been analyzed. An electrical service department competent in all its skills is far better than jacks-of-all-trades who might hide the gaps in their knowledge; gaps which could lead to serious accidents.

A second similar example concerns production and maintenance jacks-of-all-trades; a group that is becoming more widespread in France because it permits a wider range of tasks and frees operational personnel for other work. Appendix 1 discusses the insertion of such a group in an oil company. Such groups cannot be grafted on without many precautions being taken—and often cannot be applied at all during the initial years.

A company is a living organism, experiencing as such an evolution; but it is significantly different from other living beings because it changes its structure and experiences other mutations without passing through reproduction or successive generations, as with other species.

Company evolution passes through three stages: differentiation, integration, and common place acceptance. It is a natural order and can hardly be prevented where a grafting from one company to another is involved.

Adaptation of the Microstructure—Job Creation

A French uranium processing plant, the reference plant, has a manufacturing staff composed of a group head, an assistant, and three operators, responsible respectively for mineral feed, cleaning, and solvents. Is it possible to create an organizational structure suitable to conditions in Niger, using local labor, without damaging the mine? Dividing the processes into different sections creates six jobs: crushing, grinding, acid treatment, washing, solvents, drying. Moreover, bringing in intermediaries, such as assistant operators and a section head, creates a natural career channel leading to the job of technical mastery officer.

> *The technology transfer will be made easier by improving the organization during application: several career paths link up with the jobs. These jobs are classified by the complexity of the situations to be mastered by the worker. In such career paths, a job is bracketed by a job which precedes and one which follows it.*

Naturally, it would be easy enough to put men with little industrial experience through a long training leading to high-level technical responsibilities. But that route leads to unnecessary difficulties. It is infinitely simpler to adapt the jobs and arrange them like a ladder, each step a move upward in the difficulty of the situations to be overcome and the jobs to be mastered.

We discovered that approach in 1956, rather by accident, when drawing up the organization charts for the first turbo-alternator thermal power stations for Electricité de France, the French national electricity producer and distributor.

The charts were necessary; but also foreseen was the redeployment of personnel from the old power stations to the new. With five sections being brought into service each year, 200 people were involved annually.

In the old plants it was customary to designate operators by their technique: turbine operators, stokers, electricians. The shift foreman over them was as irreplaceable as he was difficult to train. He was a man with ten or more years experience, who had progressed beyond his basic technical training. (All training plans reveal the bottleneck in training shift foremen.)

However, technological evolution tended towards a new kind of organization. In the new power stations each boiler-turbine-alternator-transfer line is isolated from its neighbors. The horizontal links between the machines, which characterized the old power stations, disappeared. It was replaced by vertical liaison, stressing the section's continuity and autonomy and run from a central control room.

The structure thus evolved naturally, with two machine supervisors looking after the machines through direct physical contact, and a department head controlling the equipment from a distance and in telephone contact with the supervisors. The shift foreman supervises the four operating crews in a four-sector power station.

From the new organization emerged a natural self-education process: through contact with the machines, the supervisor understands them and their relation with other units better. It will be easy for him to become a department head, because when he gets to the level of an abstract control board, he will already have an intimate concrete knowledge of the operation. As for the present section head, he is preparing himself for running four units.

So a problem ingrained in the old system was satisfactorily solved with no need for specialized training.

We soon realized the breadth of its application. We have used it in a gas liquefaction plant to train operators and adjustors (see Appendix 6), in a mining complex to train mechanical shovel operators (see Appendix 4) and also in France to aid flexibility in organizational charts and avoid "competence barriers." For example, putting a manager into a minor accounting job prepares him for a higher job demanding deeper accounting knowledge.

We call such organization "self-training." We recognize the self-educational possibilities of an organizational structure as being an indication of its ability to develop its employees. It is tempting, and frequently achievable, to construct a link between self-educating jobs and real situations as described a bit later in this chapter.

Recruitment

Recruitment during a technology transfer takes many different forms. It may be a question of recruiting either from inside or from outside the receiver company the employees to run a plant or service department (sales, accounting, personnel). The choice will be made among those suited to the new jobs as defined in the policy criteria laid down in the preliminary study. For example, priority may be given to local labor from a plant which is closing down, to certain nationalities, or to workers who have received a certain kind of training.

But it may be preparation for a new technology in all, or part of, the company. In that case those people affected by it must be identified, their present educational level analyzed and understood, and their attitudes to the change foreseen. The most favorable and the most hostile men must be identified.

In each case recruitment has a double target. One is understanding the situation, basic education, and professional attitudes; the other is to detect where development can most easily occur.

This operation is not divorced from the other key transfer operations. For that reason the choice of the system's complexity level (*S*) and the method of assembling jobs (*O*) take into account the capabilities of the men engaged, while the training method (*T*) will be based on their educational level. Proper employee selection in the right conditions improves training efficiency and can cut the time needed to adapt trained employees to the plant.

In practice all recruitment operations use more or less sophisticated tools that deliver fairly precise assessments and classifications of how quickly the existing population, or those to be recruited, can assimilate the technology. But the problem is rendered more difficult by the fact that the tools, such as tests, are often drawn up for people with an industrial background; the results are often judged in the light of that background.

In his book *La Promotion Technique du Travailleur Analphabete (Technical Advancement of the Illiterate Worker),* Charles Maguerez reports an experiment using the Piorkowski test among miners from the Moroccan Atlas mountains and refinery workers in the Sahara. The object in both cases was to find among a large number of illiterate workers, those most suited to higher company jobs through further training, which combined technical and literacy skills.

The Piorkowski test (see Figure 4-1) consists of four vertical rods of unequal length fixed into a circular base, plus a number of similar discs with six holes around their edges. Only four of the holes correspond to the positioning of the rods. The examinee threads the rods through the right pattern of holes in the discs. That means that each time he must match the holes with the four rods. That operation contains three phases: comprehension, learning, efficiency; the test appears well suited to the problem we faced.

Unfortunately, in use, the examinees, who were used to working with heavy, crude tools, were evidently incapable of manipulating the small test discs which

Figure 4-1. The enlarged Piorkowski test.

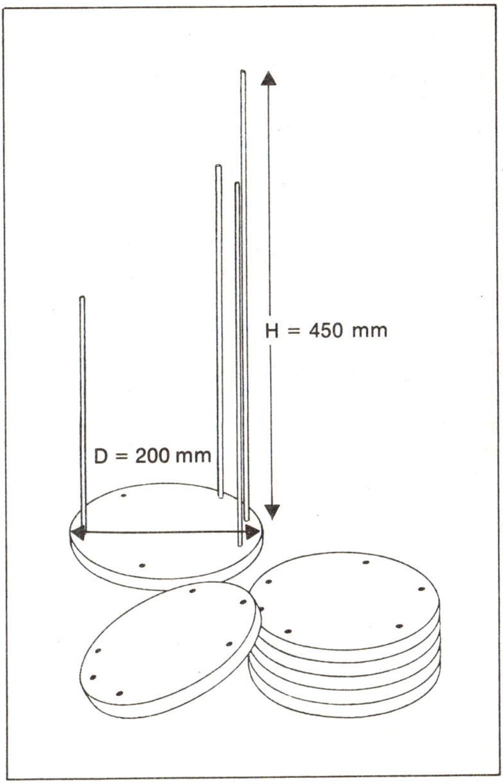

measured only 20 mm in diameter. The narrow diameter of the discs, rods, and holes deterred the candidates. However, we were not seeking precision workers; on the contrary they were wanted for careers in heavy industry. The solution was to devise the special PG test (PG for Piorkowski Giant test); the discs and rods were made ten times bigger. Thus, the rods were up to 450 mm long, the diameter of the discs 200 mm—dimensions perfectly suited to the dexterity of the examinees.

That example reveals the problems of grading and measurement, the importance of external test factors, as well as the different ways people adapt to tests; all can distort the results. And while external factors are negligible for some workers, they can be of major importance for others. The test must therefore consider the objective behavioral factors as much as the education and attitudes of the men examined.

Recruitment's goal, based on certain policy choices bearing on the workers to be selected, is to discover the best-equipped workers for the jobs and to understand the culture of the those whose jobs will be influenced by the new technology.

It centers on a study of the workers involved in test situations, who must be chosen by finding the closest link between aptitude and career destination. The grading of these situations deserves the closest possible attention, because of the unique worker/job configurations.

Leaving aside the people for whom the Piorkowski test is designed, we will cite another actual case. The problem was to find among a company's executives, the men most capable of initiating a strategy of change. Should we choose those who had already received training in company strategy in a European or American management school? Certainly not. That would be like choosing military strategists only from among those soldiers who had attended military schools, while in fact history shows that great generals reveal themselves solely through action. Another solution was to use psychological studies in leadership.

However, we chose a different method. It consisted of creating and observing an actual situation that satisfactorily represented the target situation. We called in groups of five or six top managers who had shown their qualities as strategists and agents of change. Each presented to the group an actual successful case history involving change resulting from his personal initiative. During the presentation a psychological portrait of the progress-oriented manager, was drawn up by Dr. Marcel Filiu, doctor of medicine, neuropsychiatrist, in charge of applied human sciences at Eurequip, from observation of the dynamic interplay among them. From that portrait it was possible to select executives most open to strategic planning and the company changes it implied.

A similar study was undertaken for control room operators. It began by observing experienced and efficient operators. From that study resulted a portrait of a qualified operator, followed by the personnel tests that permitted the selection of future operators.

Training

Real Situation in a Real Environment

In 1960 the Moroccan finance ministry sought further training for newly-promoted rural tax inspectors and controllers. There were 100 of them, average age 35, with long professional experience. All reacted negatively to the prospect of returning to school. Faced with that reaction we abandoned the classical teaching methods for methods based on concrete situations that the men would come across in their working lives.

The situations were frequently taken from the trainees' own experience. The cases started simply, but got progressively more difficult, demanding more knowledge as they progressed. But the fact that learning should be based on experience was a basic necessity. The trainees adapted to the courses with no difficulty and made progress in mastering more and more complex problems as their learning base developed.

> *The traditional way of training young people has three main characteristics: separation of the disciplines; advancement in each discipline by abstract and general steps; control by examinations unrelated to actual situations.*
>
> *Yet for the already-working adult that path is less rewarding (and in any case less motivating) than taking a more original approach characterized by the mastering of concrete situations. As in real life, those situations will be interdisciplinary and get more difficult as they go on.*

This refusal of the "one best way" can be substantiated by arguments developed earlier in this chapter under "Selection, Training and Role of Technology Transfer Leaders." But rather than go back to that point it seems preferable to compare the various kinds of education suitable for an adult and a child. Where the latter is concerned, we report on it as it exists, having no special experience there.

The child allows, or tolerates, a course of study being imposed on him in which he has no immediate interest. It readily distinguishes between the important world of play and the world of school.

The adult is handed a role by society and adjusts himself to learning what is required to fill it. Knowledge without a purpose deters the adult particularly if it is presented to him in a way different from its original, practical discovery. The African peasant, for example, would be very interested in the Egyptian geometrical system of measuring land, but is put off when it is presented in the logical-deductive method so dear to mathematicians. It must be added that a laborer's educational curriculum in an advanced industrial nation goes far beyond what is really necessary for his job. And we are the first to encourage, and set in motion, ways of enriching jobs so that an individual's education expectations are fulfilled.

On the other hand, it would be wrong to recommend the same volume of general education that an industrialized nation's workers receive to a nation racing against time to industrialize. A toolmaker's professional certificate at an elementary level is a promising base on which to build training applicable to most operations, maintenance and trades in industry. It was from that base, for example that we trained the manufacturing and maintenance staff for the Fos-sur-Mer refinery (see Appendixes 1 and 6). It proved possible and rewarding for both employees and company; but such a base is in no way an absolute necessity.

While a plant is being built, would it be reasonable to impose a similar training course on local labor—often adults with no technical background? Such a course would condemn us to arrive when the battle was over; the plant would be finished before the training and few of the intended workers would remain.

Adults prefer to be in contact with the real world, dealing with real situations, advancing from daily realities and treated in a real organization or at least in the human work environment. And that is just as true for the manager in a large company who is interested in data processing because it can help him with daily problems, as for the nomadic shepherd who is interested in mechanics because it applies to a tractor or a plough. And the man who knows how to drive a tractor is on the way to mastering more complex machines: draglines, locomotives, and even power stations.

The introduction to real situations, which is the basis of adult education, can take two forms, depending on the objectives: (1) training in a group composed of trainees destined for the same job, or different jobs in a common framework; and (2) group training of men from the same organization, training in different jobs, but which complement each other in operation.

The first is dealt with in "Detailed Educational Programming—The Complement to Practical Training" and the second is covered in "Group Training." Figure 4-2 represents two kinds of training introduction; it distinguishes between horizontal training, that is by level, and vertical group training, that is by organizational unit.

Detailed Educational Programming—The Complement to Practical Training.

When it began, the Miferma operation was composed of one instructor for each group of ten trainees, who prepared and carried out a three-month training course: eighty instructors were needed at short notice—where were they to be found? The fastest solution was to find them among the future supervisors, who would be recruited one year before start-up.

The gaps in that solution are self-evident: those men had no experience of teaching, or of Mauretania. Moreover, their knowledge of their trade was frequently not logically thought out and therefore difficult to transmit.

The gaps had to be filled, or the instructors would not be correctly performing their jobs as technical assistants. And it is better to do it before start-up by giving them responsibility for training. But on condition that each instructor was backed up with teaching methods clearly adapted to the situation.

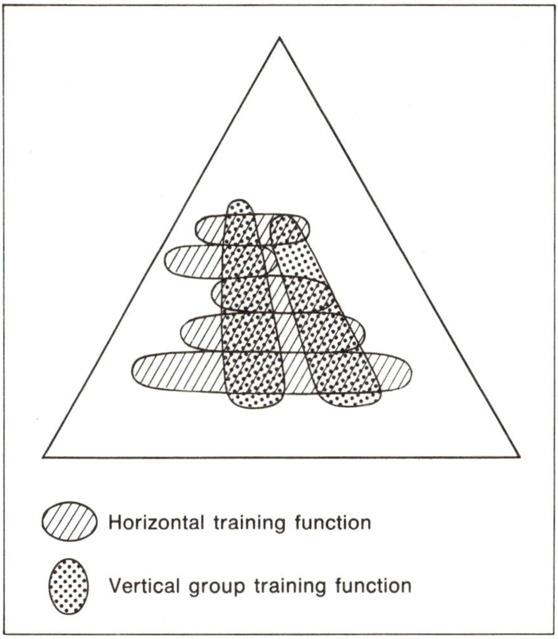

Figure 4-2. A company organizational chart showing the training function; men of the same level (horizontal) are brought together and trained in groups drawn from an actual working unit in the organization (vertical).

A common introductory course at the plant and at Miferma started each training period, followed by, in some cases, a common mechanical electrician's course, completed by a short specific job course. The large number of people taking the general courses justified intensive preparation and the manufacture of appropriate educational material.

Each instructor was thus guided as efficiently as an actor following a text and a scenario.

> *Practical education should avoid improvised and haphazard training. On the contrary it demands much thought about the choice of educational methods, taking into account cost, duration and effectiveness; and a careful study of the teaching program, as it will be taught by nonprofessional instructors taken from the staff closest to the jobs demanded.*

There is a danger that these solutions will drift towards improvisation—it usually accompanies on-the-job training. But the proposed method steers clear of that. And in fact the operation's success is linked with the detailed blueprinting of various operations running concurrently with their exact definition and the allocation of resources for success.

It would be useful to dwell a moment on what is meant by training operations.

Technology transfer, as we have already stated, is a group business. The transfer will be successful if we transmit to each cell in the groups the behavioral, organizational, and educational techniques sufficient for its new job. We know it is possible to analyze behavior according to methods laid down by the educational program specialists, which describe today's behavior among each population and the behavior they must acquire.

To make the transfer succeed the consultant can use a number of means, some already existing, some to be created:

1. School, plants, assembly shops, which can receive trainees
2. The training "components," lessons or educational material to be used
3. New courses adapted to different staff levels
4. Training material that must be specially created and can consist of audio-visual aids, charts, simulators, television, programmed books, etc.

Each of these methods has its particular costs, setting-up periods, probable efficiency, and linkage with other methods.

Like an architect whose job touches several different trades, the transfer consultant must propose the combination of methods best suited to time and cost criteria. If, for example, it involves accounting for managers, the consultant may hesitate between a programmed teaching course and one using teachers. The cost reasoning guiding his choice is illustrated in Figure 4-3.

The preparatory investment is higher for the book than the teacher (by 1 to 8), but the preparatory cost is higher in the second case (the teacher's salary) than in the former (purchase of additional books). That example demonstrates that programmed book teaching is a better cost choice when the students begin to number 1000. Similar calculations are possible for other kinds of training such as workshop schools, video cassettes, simulators, and so on. The results, more valid because they contain time and effectiveness criteria, allow the consultant to establish a list of alternative solutions.

The consultant will pay particular attention to made-to-measure educational programs where they appear inevitable, but will first have examined the range of courses available and made choices among them. The following two paragraphs are devoted to the creation of made-to-measure programs.

Macroteaching: Orientation and Program

During the remodelling of the professional aptitude certificate program at a technical institute of public works a question arose: how does one teach basic mathematics—indispensable for public works professionals—to young people

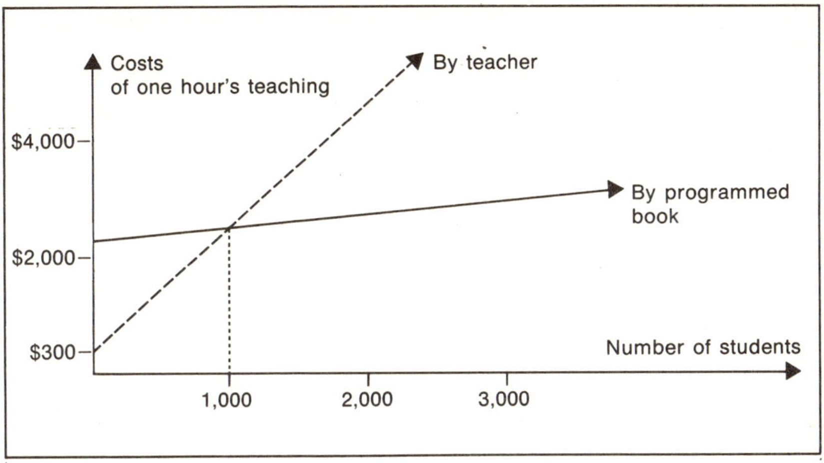

Figure 4-3. Comparison of one hour's teaching costs.

sent into that career after a short training taking into account their mathematical inaptitude?

The managers responsible for the program found inspiration in a similar example: the organizers took Swiss Family Robinson as their theme, imagining their pupils shipwrecked on a desert island. To survive they had to build a hut and plough, and cultivate and irrigate fields, retracing the slow progress of their forebears. Mathematics followed as a kind of afterthought, permitting them to measure, compare and conceive. That kind of useful mathematics, born from observation, was perfectly assimilated by the pupils.

The creation of a program begins with two analytical operations—the life-style of the people to be trained—their cultural characteristics, which is the departure point for any progress; the intended professional behavior at the end of the program.

To pass from the first to the second stage, a training orientation will be selected, based on the trainee's interest which can be represented by a training tree.

We experienced many disappointments before arriving at the above rule; it appears complex, but is relatively simple in application. From the beginning we were compelled to analyze the jobs for which the program was to be prepared. Reading the works of Crowder and Skinner, pioneers of programmed teaching, directed us towards a richer concept of the job study, detailing the necessary behavior of the job holder. A job may be described in all-embracing terms such as "running a separation unit," which presupposes knowing how to run a separator, understand the instruction manual, reading operational charts, and so on.

The educationalist has complete freedom to choose the educational route by which he moves to his objectives, and has a large number of starting points to choose from. Each starting point will be marked out by a behavioral situation route that indicates the steps between the starting point and the objective. It can also be represented by a tree—similar to a decision tree—where all the ramifications will be interwoven from origins to goals, passing through the relay stages.

It is essential to respect certain rules when drawing up an educational tree:

One logical rule: learning new behavior presupposes building on behavior already learned.

Three educational rules: (1) all newly acquired skills must be quickly put into practice to ensure they are consolidated and not forgotten; (2) a new idea or skill will be much more quickly learned if put into an actual situation; and (3) a program comprising various disciplines must be marked along the way by reference to actual situations drawing on the knowledge acquired from several disciplines.

The logical rule and the first educational rule are taken directly from Skinner.

The originality of our second rule was revealed during experimentation: the adult's curiosity must be awakened by a real situation that retains his interest, and one that he understands he cannot master without certain knowledge. The adult is thus placed in a discovery situation, and he wants to learn in order to understand and master.

But having read Nietzche's *Aurore* I was tempted to be less ambitious: "When they imposed mathematics on us by force rather than making us face the despair of ignorance, and reduce our small daily lives, occupations, and what happens from morning to night in the home, the workshop, in the heavens, the countryside, to thousands of small problems—torturing, humiliating, exasperating problems—to make us realize that we need above all mathematical and mechanical knowledge, and only then to teach us our first scientific enthusiasm before the absolute logic of this knowledge. . . !" So perhaps we retain the merit of having introduced the wishes of the philosopher into reality.

We believe that our third rule is completely original. It arose during preparation of programmed teaching works in which the knowledge transmitted is cut into small items. Cutting up the program has certain advantages, but tends to tire the reader or trainee creating a dangerous obstacle to this technique's success.

It is as if the trainee had to place himself in the teacher's hands, who then leads him by a secret route until the happy arrival at the end of the course. No stopping on the way, no respite to recap and mark one's progress. However, the Gestalt theory, or theory of form, seized our attention during the preparation of these programs, leading us to an understanding of the trainee's weariness and its remedy.

There is something incomplete in this itemized teaching which works against an easy satisfaction to be gained from the arranging of bits of knowledge, and also prevents the creation of a significant whole from the sum of the parts.

Programmed teaching would thus be like art appreciation which focused on a series of colors without leading to a study of the whole painting. It is far better to space the lessons with increasingly difficult real-life situations: the sketch, the roughed-out painting and the completed work. Such is the aim of our third rule.

Microtraining: Creation of a Lesson

An important multinational bank was worried by the unrest among its (mainly female) staff, preparing documents for the computer. A psychological study discovered that the unrest sprang from the fact that the staff did not understand the system of which they were part and particularly the computer. In this case the technology transfer was divided into too small pieces, omitting the human and technical environment in each job.

A program was put together to "demythologize" the computer. The obstacles were numerous; the staff's education level was low, their technical background almost nonexistent. The training would be undertaken by other staff members. Each lesson was meticulously prepared so as to accurately situate it within the limits of their interest and understanding.

The object of each lesson was defined in a process of continual advancement, while the organization of the lessons varied between the concrete and the abstract.

The preparation of the educational scenarios was enriched by the staff's own observations during the first lessons; it was so minutely detailed that it became the basis of a book on programmed teaching.[11]

> *A lesson is a teaching period directed by a teacher. It demands a period of preparation 10 to 40 times longer than the lesson; such preparation is a normal step and the real framework of programmed tuition.*
>
> *The arch method, including alternatives between real situations, charts and samples, attracts the trainees' interest and ensures the advancement not only of knowledge but its translation into action.*

While the macroteaching is concerned with the program's overall architecture in which classes instructed by a teacher can extend over several months, the microteaching is concerned with the creation and writing of a lesson lasting one or two hours.

Whether it involves managers or operators, the class numbers between 8 and 12—preferably 8—taught by a staff member from the organization whose teaching experience may be only sketchy.

---

[11] *Initiation à l'ordinateur.* Langages de l'action, collection published by Fayard (France).

The lessons will only be effective if carefully prepared. Our practice is to prepare a training guide of between 10 and 30 pages (depending on the difficulty of the subject), for a one-hour class. It is a veritable scenario for the instructor plotting the advancement of the group, emphasizing the key questions and corresponding replies and laying down the rules for the use of teaching aids.

In fact the preparation is similar to the preparation for a programmed learning book. The only difference is that in the second case, the program must be more detailed because the trainee studies on his own without the aid of the group dynamism, or the help of a teacher, who himself is guided by a teaching plan.

We distinguish between these two kinds of programming by the initials PTI (programmed teacher instruction) and PMI (programmed machine instruction). By machine we mean a range of aids extending from the simplest, such as the programmed book, to the most complex, the computer program of televised sequences. Our experiences with both PTI and PMI have lead to a mutual enriching of both teaching methods.

PMI lead us to a more rigorous analysis of behavior and results. PTI permitted us to ascertain the effectiveness of a class through its presentation to a group; certain teaching charts were rejected, the class finding others best suited to its needs.

Finally, we arrived at an integration of PTI and PMI; PTI needing 10 to 40 hours of preparation for one hour's teaching is a step on the path to PMI which needs 100 to 400 hours preparation. Thanks to PTI we can verify the effectiveness of a teaching scenario, not moving on to detailed programming until the scenario satisfies. PTI is thus a control during the creation of PMI.

The use of PMI, which can bring in the most sophisticated teaching media such as television, is generally based on a cost choice. Other criteria such as the need for the trainee to work on his own, can lead to its use. Such is the case with further training schemes for shift foremen in continual process plants such as refineries, gas liquefaction plants, thermal and nuclear power stations, and so on.

Returning to PTI, the most frequently used, we note that the speed at which the lesson advances, a function of the group's rapidity of understanding, is not the only programming rule to observe.

We must respect another rule, already noted in Figure 2-4. That is, the systematic use of five elementary situations making up a "training arch" which, through a chart, relates the actual to the schematic. The advancement process is described in Table 4-1; it is an example of the application of the arch to the training of a mine timberman.

That switching from the actual to the conceptual, then back again through a defined stage (the mock-up, a simplified image of the actual) takes place, according to the case, one or several times a lesson. It maintains the trainees' attention and ensures the effective penetration of knowledge acquired during their professional lives.

Table 4-1
Training Arch

Theoretical Steps	Application to a Mine Timberman
Actual observation	Observation of a straight gallery
Observation of a model	Model of the straight gallery
Discussions on diagrams: approach to a new idea	—Diagram 1, plan of a straight gallery —Diagram 2, plan of a curved gallery —Differences studied
Application of the new skills on a model	Making a model of a curved gallery
Actual application	Timbering a curved gallery in the training school

Group Training

The need for group training becomes clear when considering the application of a technology to an easily identifiable body.

The most convincing example is that of operating crews in a complex modern plant. It includes different levels of responsibility: engineers, shift foremen, chief operators, control-room personnel. Each of them will be involved in a training process taking into account his basic education and the demands of his job. There would be no point, for example, in giving the same training to an engineer as to an "itinerant operator." Neither has the same view of the same events, that view point being naturally linked to their understanding and capacity to draw general conclusions.

However, having trained each member of the team through courses which brought together men of the same level, the transfer is not yet complete. The transfer is a group phenomenon and the group must be trained—that is, it must learn to function in an actual organization. And that must start before nonprogrammed incidents so that they learn in a coordinated way and not by error.

So that the best form of group training is by tackling real problems in a real organization. By real problems we mean an actual case examined, presented to the group as a case study, or (more difficult, but more rewarding) a real problem without a solution, put to the group, or better still, discovered by them.

Training is generally carried out by meetings of trainees of the same educational level. They are destined for identical or similar jobs. The juxtaposition of different levels of skills in an industrial organization will not automatically generate technical competence. The next step is the training of interstaff groups through the treatment of real problems in an actual environment.

Group training is therefore conceivable and effective in all human groups receiving a new technology. Where it enters at the beginning of a large training operation (for example the preparation of a crew for a plant under construction) it is even easier to put into place. The communication channels already exist, staff relations link with the training relations and it is quite natural that each shift foreman, for example, will benefit from incidents happening to others to find a discussion theme for his work team.

The same process arises during a strategic management application (see Appendix 7) where groups discuss and think about similar techniques which might affect the future of the units for which they are responsible.

Where the group is less easily identifiable than an operational group, training through real problems can extend through its action; it might require the presence of a manager, remotely located in the organization, who is considered necessary to open up a facet of a situation less well-known by the members already there. In this way, the life of the group approaches as nearly as possible the life of the organization in action.

The Dynamic of the Whole

A bottleneck suddenly appears around the dragline operators, slowing down the expansion of a Moroccan open-cast mine. Training is organized in a pilot yard where a self-educating system is used—lorry driver, bulldozer driver, dragline operator, site overseer. A gap exists between the bulldozer and the dragline which is bridged at the dragline school by a middle-sized bucket.

The transfer succeeds therefore, at the pilot yard, reaching its objective, a 60% output of the dragline's capacity. Does it end there? Certainly not. Three objectives remain:

1. Increase output above 60%
2. Increase the number of nationals in the other mines
3. Open up new mines

But these objectives cannot be pursued simultaneously. A transfer strategy is drawn up giving different priority to each objective. The strategy is expressed in a provisional management plan permitting a scheduling of men along the career channels defined above.

This plan progressively improves the running methods—tasks become more complicated, from which higher performance can be reached. The jobs of the expatriates evolve too as they assume a greater training role, while direct running jobs are given to nationals.

The transfer must not be conceived as a once-and-for-all operation. It will certainly be satisfactory if a certain number of targets have been reached, testifying to their arrival at an "objective" mastery of the job.

> *But the transfer advances through action, sometimes automatically, and results in a permanent improvement in the group's efficiency, which will be actually better than forecast or planned. Briefly, there must be a permanent dynamic in the transfer, which can be expressed in forward-looking men, jobs, and systems management.*

The triple dynamism of men, jobs, and systems, has already been raised in preceding pages, but is so important that it deserves to be elaborated on. The basic idea is not to oppose certain natural advancements, but, on the contrary, to use their dynamism to arrive at the development and satisfaction of the men and the group. Carried forward by training and experience in his job, the employee improves his performance over a period while developing his motivation. After which he becomes prey to routine, while habit leads to weariness and lack of interest.

There are three ways that management can avoid these three temptations and risks of getting into a rut.

The first is *promotion:* or even simply developing all-rounders or jacks-of-all-trades: it will be easier when the self-educating structures are in place, each employee preparing himself for a higher or neighboring job. In developing countries the route already led to a quantitative and qualitative change in foreign technical assistance.

The second is job *enrichment.* Without changing the system it regroups the tasks, leaving greater responsibility for each employee depending on his maturity. It bears on the operation "O." This second course is less complicated than it first appears and is often used by top executives through delegation to their subordinates. Of course delegation can also become frozen, administrative, defining each person's domain without possibility of change. But it can also promote a manager's development, when his superior gives him progressively greater operational responsibility in his field.

The third, bearing on operation "S," *is progress in the technical system.* This is never perfect nor unique—and that applies to every aspect of the company. It is interesting to establish a dialectic progression between the group's educational level and the level of complexity of the systems it has mastered. This is never easy to do since it involves machines, and must be considered a fixed condition. But a factory operational system, and even more so with the maintenance or control systems, can be more and more complex without the need for machinery.

That range of choices is illustrated in maintenance. One can be quite happy with an elementary system, where repairs are carried out when accidents happen. Yet one can also aim higher and prevent the accidents by inspections based on simple criteria such as the number of running hours, or the quantities produced. We see immediately that such a system is very simple indeed, which leads us to seek more significant and complex criteria to follow. That is the road to forecast maintenance, based on production curves, temperatures, pressures, vibrations and so on, without stopping the machine.

Each time the system grows more complex, more difficult jobs appear which cannot be undertaken if the group is not sufficiently educated, but which, on the other hand, permits a qualified group to be identified. According to the cases one might prefer to forecast and organize the future management of the men, jobs, and systems; or to leave them a certain freedom to evolve, which can, in extreme cases, lead to free division of tasks in an autonomous group.

Measuring Results and Corrective Action

Before the start-up of a petrochemical plant the operating crews were trained on life-sized wood and cardboard mock-ups of the operations room and control boards. Photographs took the place of machinery and simple accessories changed instruction, dial readings, and graph tracings. In an environment very similar to a real one, the mock-up made it possible to represent any kind of situation; it can function like a simulator with the instructor at the central unit backed with fictitious or real case histories from similar plants.

The object is to make the crews capable of mastering increasingly difficult situations. Two stages can lead to that goal:

1. The teaching tool, that is, the simulator and the case histories, must be adapted to the knowledge, and the capacity of the audience to structure its thoughts.
2. Each member of the audience must pay close attention to get the best out of each class.

Two controls are necessary which are similar, in industrial production, to production tool quality (proof of tool receipt) and the quality of products during and after manufacture. The first control is statistical: one aims for a certain number of correct replies from the class at the end of the lesson. If the right number is not arrived at the educational tool must be changed. The second control consists of checking trainees' progress during and at the end of the course. Whenever anomalies arise, particular attention must be given to it. Is it an orientation error? Has the trainee lost track at a certain stage? Can he catch up with the class by making an extra effort? In fact the same test can be used for both control operations; the results appear in a column for judging lesson quality, and on a line for judging trainee progress (see Table 4-2).

Yet sometimes the results may be unsatisfactory despite an improvement in the educational tools. In that case we must go upstream of the simple training operation, examining on the one hand, the suitability of the organization and even the conception of the system, and recruitment on the other. Experience shows that small improvements in the details of these operations (for example, handing back to management jobs which are judged to be difficult, or the introduction of a new selection tool) can exert a great influence on the quality of results.

Table 4-2
Checking Teaching Effectiveness

Classes / Trainees	Procedures Learned	Measuring Studied	Instrument's Significance	Normal Start-up	Emergency Stop	
Andrew	+	+	+	−	+	
Bruce	+	+	+	+	+	
Charles	+	−	−	−	−	← trainee to be reexamined
David	+	+	+	−	+	
Frank	−	+	+	+	+	
Lewis	+	+	+	+	+	
Matthew	+	+	−	−	+	
Paul	+	+	+	−	+	
Robert	+	+	+	−	−	
Victor	+	+	−	−	+	
			↑ Ideas to be revised	↑ lesson to be given again	↑ Ideas to be revised	

The group education and training stage is the time to introduce the most significant controls in the man-job, group-mastery technique.

Two controls should be frequently applied, in principle, either after each lesson or each day. The same tests can be used in each case.

The first centers on the relating of the educational tools to the aims of each lesson and the audiences' characteristics. It also permits a judgment on the operations leading up to the creation of the system and organization, on the one hand, recruitment on the other. The second judgment is made on the actual progress of each trainee.

It is, moreover, advisable to plan an overall control at the end of the training period, to verify the effective mastery, by each man and then in each group, of the professional aptitudes which the program aimed at.

A rigorously business-like approach in the control of results is essential—the sole criterion is efficiency. Results must be as close to 100% as possible. So how is it done?

We must carefully distinguish between the tests at the end of the training (for example, before trainees take up their jobs) and the controls on the efficiency of the educational tools and the progress of each trainee. The latter controls must be frequently applied—sometimes after each lesson. Their aim is to

1. Verify the suitability of the educational tool and refine it wherever the result is below 90%

2. Where results do not improve, go back to the organization and conception of the system to improve them where necessary
3. Identify trainees in danger of dropping out, so that they can be pulled in, in time

In practice, the aims of each lesson are test-controlled—preferably before the lesson is prepared.

Examination should be conducted in the same spirit—that is, to say, relating them not in relation to a theoretical program, but to the real job for which they were prepared.

This permanent control on efficiency must not be restricted to the training period. When the trainees are in their jobs, they must be closely observed in the different situations they were trained for so as to remedy any errors that may appear. This essential on-the-job control will be most easily carried out by the staff instructor.

The Transfer Strategy and Creation of the Overall Plan

The technology transfer takes the form of a project, often a big one, in which apparent complexity hides a large number of simple sequential operations.

The creation and start-up of the operations in an unordered fashion would be like starting work on a building without an overall plan.

A technology transfer plan must always be prepared. It should lay down the itineraries for selection, recruitment, training received or given, organizational work and so on, to be followed by each person and each group concerned by the transfer.

A technology transfer usually has no simple solution, nor unobtainable solution.

No simple solution: the problem will not be solved simply by setting up a training center, or sending the people involved overseas after a training period in a similar plant.

No unobtainable solution: provided one takes action in time—that is, at the beginning of the course where the work crew for a plant under construction is involved—to admit the need for specialized technical help at the beginning. And for a period running from a few months to several years, a pluralist solution involving various existing methods or methods to be defined and realized for the particular circumstance, can be conceived and carried out successfully.

That is the real dimension of a technology transfer. A large project involving hundreds of people is like a vast military operation which breaks down into hundreds, and sometimes thousands, of small operations. Each of these simple operations must be foreseen in detail and fitted into their place in the chain of events.

However, this approach to the logical sequence of tactical operations (that is, the transfer strategy) will not always take the same form. As noted earlier if the group's background is sufficiently developed it is possible to give them the freedom to decide the various transfer operations sequence. The classic "one best way" must be ignored when we are drawing up the itinerary for a group seeking industrial skills. The *T* phase may well be inverted, this primary phase leading to successive phases: that is, the natural dynamism and spontaneous development of the group.

As is often the case for the preparation of the work crew for a plant under construction, a convenient way of analyzing the overall project is to define (for each job and origin of the worker) the path to be followed by each man so that he is ready to start at a given date, within, and in relation to the management structure of his group. The juxtaposition of the various schedules reveals the obligations to be respected in the organization of the operations.

Some of the jobs seem complicated, demanding long experience. How do we draw that job profile for the appropriate post? How long does it take to bring a man of a certain education level up to a certain industrial level? What specialists are needed for the preparation of the necessary training? How long will it take? Lacking a time estimate the answers can only be found by referring to actual case histories.

In other words the drawing up of the plan is the key to a technology transfer. And the ability of the transfer consultant to successfully carry out a similar operation in reasonable time appears to be the least pledge he can give of his effective professional maturity.

5

Upstream from the Transfer: Creating the Teaching System

Preparing a workforce for a plant under construction is a special technology transfer case, but also a limited one. First, because of its urgency, the need to respect the completion deadline, which is rarely longer than two or three years. Second, because of the evident need to tap labor sources, such as they are, from an educational system, which there is no question of modifying at such short notice.

Yet that limit has its virtues. Mobilizing people's energies, it aids the introduction of unusual, but effective, training methods. And it starts management thinking about the structure of local technical training, which frequently seems inappropriate to the actual problem, but which must still be solved.

Fortunately, we can escape some of the handicaps of this last-minute race against the clock by paying special attention to the educational system, and trying to set up at least part of the system as an advanced guard for a vast technology transfer network.

What Is a Teaching System?

Even though the term "teaching system" is frequently used, it is not so well established that it can pass without certain preliminary explanations.

A teaching system is one or more establishments—training centers, schools—that are linked by a common feature, such as a similar goal. But why then the use of this ambitious and ambiguous word system? Simply to underline, as is frequently necessary, that a school is not just new buildings, but people, masters and pupils, educational programs, equipment, and procedures, all working towards the same goal: to lead the pupils to a certain level of knowledge and skills. (Traditional school buildings are not necessary; there can be old buildings, barracks, camping sites, worker's huts, etc....or even do without buildings and mix training with action in the work place.)

However, the goal may be foreign to any technology transfer aim. It may be purely the pursuit of knowledge, wanting nothing to do with the world of action, denying that it is preparing people for the working world, admitting only reluctantly that it is ensuring its continuity by creating more teachers. Such a system appears to be outside the scope of this book.

And yet, as is frequent and even widespread in companies and trades in developing and developed countries, the aim of a teaching system renders the system an essential part of one or more technology transfer systems. That aim can be either technical mastery, or, more frequently, a milestone situation leading to mastery in several techniques. We restrict our field of thought to systems with such aims, which we might call "professional teaching systems."

But to justify such a term, the system must envisage a certain continuity. Thus, a training system created specially for a plant start-up is usually not a teaching system. It is, rather, a training workshop, destined to disappear like other workshops, or to continue in a reduced form.

The opposite of a teaching system is, for example, the Algerian technology institutes or even the training system for nuclear power station managers, set up by Electricité de France, the French national electric company.

Why Create a New System?

New and Important Differences Appear at Different Levels

It is increasingly frequent that the various economic and social organisms (which can, in rising order of importance, be a public or private enterprise, a professional federation, or even a nation) find themselves needing personnel educated to a certain level, but their medium and long-term availability is not sufficient, so a specific, ongoing solution is demanded.

A detailed analysis of the circumstances surrounding the creation of a new system reveals the five following criteria, to which must be added a nonsatisfied need.

α—Specific characteristics of the knowledge to be transmitted
β—Human or economic responsibilities of those to be trained
γ—The people involved
δ—Forecastable duration of the need
ϵ—The wish to adapt a certain category of people to the needs of the labor market.

The significant decision criteria are not always the same, depending on the level of the principal actor in the operation. This interdependence between the actor, operation, and criterion is expounded in Table 5-1 where each criterion is represented by the following value symbols:

weak +
medium + +
strong + + +

Table 5-1
Creating a Training System Relationship Between Actors, Kinds of Training and Decision Criteria

Actors	Kind of Training	Decision Criteria				
		α	β	γ	δ	ϵ
Company	Training school			+++	+++	++
	Management improvement	++	+++	+	+++	
	Specific job training	+++	+++	+		
Branch	Preparing men for similar jobs in different companies	++	++	+++	+++	
Country	Preparing men for different technologies (common skills)	+ to +++	+ to +++	+++	+++	+++
		(according to the levels involved)				

α—Specific features of the knowledge to be transmitted

β—Economic or human responsibilities of the men to be trained

γ—People involved

δ—Period of the need

ϵ—Wish to adapt a certain category of the population to market needs

They are dispersed among the three principal actors as follows:

A large company. We know that a number of companies in developed or developing countries have various training systems, often well-structured and comprising:

1. Apprenticeship schools (thus the importance of γ δ ϵ)
2. Systems for further management training (α β δ)
3. Systems for training high-level specialists, directly related to company needs (β)

With regard to this last example we can cite the training of light-water reactor operators for important electrical companies such as Electricité de France, or again familiarization with a certain kind of airplane for a large airline, such as the Boeing 747.

In both examples relatively few people are involved (γ); the need is linked to the predictable life of the technology, not to its indefinite continuation (δ); but the knowledge and skills are highly specific (α) and emphasize again the economic and human responsibilities (β).

A federation or industrial branch. The example of a technical institute of public works illustrates the kind of problem (specific to an industry, but not to a company) which can be handled at federation level; α and β have therefore a medium value, while γ and δ a higher value.

A nation. As the actor's responsibilities rise, so do the problems become more widespread, and so the criteria γ, δ and ϵ tend to preponderate. At the same time, this rise in the actors level from the particular (the company) to the general (industrial branch, then nation) give rise to two new constraints—diversity and development.

Diversity means that if we are interested in the needs of a nation as opposed to a company, then it is not possible to plan training fitted directly to a specific job. Such a training system cannot therefore really be considered a technology transfer, which is most often encountered where men must be trained for specific jobs in a company. And where the company used identical equipment, whether Boeing 747s or 900 megawatt light-water nuclear power stations, the training system naturally covers the whole of the "individual training" component in the transfer system. In order to bring the trainees up to the required level of competence, all that is further needed is group training in the company.

On the other hand, when the jobs are no longer identical—which is often the case with a company (a refinery, for example, with widely differing functions), always the case with a federation (to the plant diversity must be added structural diversity, that is, the workers' jobs), and even more so with a nation—the teaching system must accept and even use that diversity. It must go beyond the preparation of workers for specific short-term job needs and give them the capacity to adapt to varied immediate and long-term situations.

The long term leads us to the second constraint—economic development, a region or country's long-term economic development (usually contained in a development plan) cannot be forecast to the point of defining precisely the jobs needed over a long period.

In this case the aims of the teaching system differ from those of a technology transfer. The aims of the first come before those of the second; and if the analogy is not too repugnant by being too schematized, we can compare the first to "semi-finished products," the second to "finished products."

But to return to the circumstances in which the new needs appear, the situation clearly differs according to the development level of the country concerned.

Let us start with an industrially developed country. It has a heavy professional teaching infrastructure at various levels which is gradually veering away from its traditional industrial users.

A general rule about the gradual development of the educational structure can be drawn from observing the birth of a teaching system around a need through a related structure that perceived the need (example: creation of engineering schools in various European countries), then seeing that system develop and operate under its own power by freeing itself from the originator and being absorbed by a centralized organization such as the Ministry of Education in France, or in Switzerland the Federal Polytechnical Institute which several years ago absorbed the Polytechnical Institute of Lausanne University

In either event the results are rarely contested: the teaching systems of developed countries have little connection with the pupils' future employers and turn out a finished product having little to do with the actual or foreseeable market circumstances.

The inadequacy between education and employment needs is well-enough recognized at this level by the appropriate authorities, so we will not further discuss it.

Noting the structural duality between employment and education, we can be justifiably surprised that in real life they do not match up too badly. And we might ask the naive question: "Why does it finally work?"

The reply seems simple: it works for two reasons. First, because it is a question of managing an abundance of knowledge, not a dearth; the development of the teaching level generally anticipates the economy's real needs. One example illustrates this discovery: unemployed elderly managers, which would be unimaginable if there were a dearth of management talent.

Secondly, it works because the industrialized countries possess an enormous capacity for absorption and adaptation among their public and private enterprises. There too, there is an overabundance and a rich human environment far beyond the basic needs, which helps the theoretician become a practitioner. And finally, organized systems often exist to aid adaptation to operational responsibilities.

If we now broach the case of a developing country it is to discover a totally different situation: there it is a question of managing dearth, not abundance. That difference alone justifies the search for methods adapted to their needs; and it excludes the mere repeating of industrialized countries' systems.

We must recognize that the dearth is widespread. The corrective personnel who could aid the professional training of a non-practical theorist are not usually found in these companies. A young engineer in a developed country can rely on foremen who will catch in time the inevitable mistakes he makes during the early stages of his career. That protecting envelope of experienced foremen is often missing in a developing country and the young engineer's job is therefore much more difficult, calling on a detailed knowledge of elementary technologies, rather than mathematical-economic speculation.

In such a dearth, and putting aside the needs to be satisfied, there is no question of setting up teaching systems whose sole aim would be to generally raise the level of knowledge in a country. That would restrict them to the importation of specialists and graduate unemployment.

What then is the primary object of a teaching system? According to the case, we can seek to fill certain gaps at various levels in the economy, to ensure the replacement of foreign managers, or to provide the jobs forecast for the medium term.

The third hypothesis is clearly the most fertile, it will be the most frequently encountered when the catching-up process is over. But it is in no way easier. It is a question of providing jobs forecast for the medium term and of creating a teaching system that anticipates economic development. It is tempting in such circumstances to pay greater attention to the often-critical short term, rather than the medium or long term. One might, for example, be tempted to create a permanent teaching system centered on the near-immediate needs of a plant being built.

But that step presents certain inconveniences. On the one hand, the creation and application of a teaching system contains greater inertia than a technology transfer system. The study period is longer because the field is wider. And we run the risk of arriving after the battle is over, that is to say, to arrive with the work crews a long time after the plant start-up. On the other hand, the tree, which is the plant under construction, blocks the view of the forest, which consists of plants being planned. By being too-well adapted to qualitative and quantitative short-term needs, the system runs the risk of not being suitable to the medium-term needs, which it is nonetheless its job to fulfill.

When drawing up a teaching system, it is therefore advisable to be thinking about the projects five or ten years away, rather than solely the work in hand.

A new teaching system must satisfy the organization's development plan and be conceived synergetically with it.

Synergy in this instance means that the system supplies the workers needed in the plan, while the plan provides jobs into which the trained workers—sometimes, if necessary, after further training—can be professionally integrated and consolidate their knowledge through practice.

The Carbon-Copy Transfer of Existing Systems Proves Unrealistic, or Impossible.

The preceding section ran through the various circumstances involved in creating a teaching system. This section will show that the needs rarely fit in, purely and simply, with the transfer of an existing system.

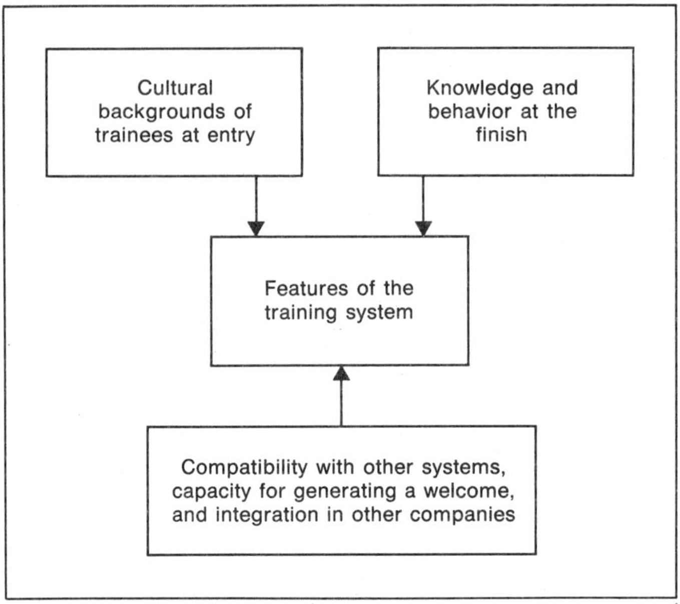

Figure 5-1. Factors that influence the conception of a training system.

Figure 5-1 charts the major influences on the creation of a teaching system:

1. The educational level of pupils at the entry stage, frequently decided according to the people available rather than the real demands of the economy.
2. The knowledge and skills necessary at the end of the course, more or less linked to the development plan, according to the level of the system's actor.
3. A certain coherence with other teaching systems (it is frequently sought and always exists), which are conceived to create new management posts and personnel capable of generating companies' ability to receive and integrate technology.

Besides these three influences, which are clearly specific to a situation and identically reproduced, we can mention as a further transfer difficulty, the degree of organization of the system we wish to reproduce.

Lowel Steel's remark in Chapter 3 while relevant to an industrial environment is even more pronounced in an educational environment: teaching systems that go beyond company training are rarely well-enough organized to be easily reproduced. Teachers' traditional independence of spirit is accompanied by an almost complete absence of documentation.

All carbon-copy transfers must begin by assembling all existing data; that requires an effort almost as great as the creation of a teaching system perfectly suited to local conditions.

It is far better, therefore, to exclude identical transfers; but that does not mean that we exclude the reuse of certain teaching components taken from one or more of the reference schools.

How Do We Create a New System?

The transformation, creation, and application of teaching systems respond to a certain number of imperatives. Each imperative is a step towards modifying the existing, or implementing and operating the new, training organization. Of course, the fields involved in the various stages will differ according to the principal actor, industrial branch, or nation.

Yet, despite the differences revealed above, contrasting the teaching and transfer systems we still find many similarities between the key operations in both cases.

Seek and Define the Objectives. There must be a coherence between the teaching system and socio-economic tissue into which it will be placed. That coherence expresses itself in qualitative and quantitative objectives.

Qualitative Objectives—What will the workers' missions and jobs be when they start their professional lives? Where semi skilled workers are concerned, the jobs study is "normative" that is, it must take into account the developments that are likely to affect their jobs in the medium term, insofar as it is possible to define and sketch their outlines in the initial study.

Where managers are concerned the jobs study consists of our considering: (1) the professional situations they will experience; (2) the work methods to be used; (3) their professional tools, such as mathematics and so on; (4) the necessary data, particularly technological; and (5) the qualities, attitudes, or aptitudes to be developed.

Quantitative Objectives—What will the needs be for a given economic sector, region, or country at large from the viewpoint of expected development? This kind of study is carried out by a checklist taking into account the evolutionary, quantitative forecasts, or from the basis of a development plan.

Study the Population (see "Preliminary Studies," Chapter 4). It is from this source that the trainees will be taken. This study, which must also be quantitative and qualitative, is indispensable, but frequently neglected. Yet, it is necessary in order to evaluate the training time compared with the starting point, the level to be attained, and the adaption of the teaching methods.

Establish a Needs Calendar (see "Preliminary Studies," Chapter 4). The lack of competent personnel at the right moment holds up economic development. Consequently, the qualitative and quantitative needs defined above must be managed so that supply and demand are in harmony.

It is a question of aiming accurately at, while taking into account as much as possible, the further training vital to "semi-finished products" coming out of the teaching system so as to be ready to assume their responsibilities at the right moment.

Study the Existing Training Organization (see "Preliminary Studies," and "Selection, Training, and Role of Technology Transfer Leaders," Chapter 4). When the needs are clear, and before creating the training system, it is important to draw up a detailed educational organization list (in schools, companies, etc.). It must cover existing and potential organizations within the economic sector, region or country, which can be used directly or indirectly, totally or partially.

A concern to limit equipment or intellectual investment while aiming for the objectives will be reflected in the cost of teaching.

The use of existing sources of knowledge, notably locating reference schools and finding useable educational components, and resources for men in the industrial branch concerned with transmitting knowledge of industrial training structures are major assets that the creator of the teaching system must have in hand.

Choose the Leaders (see "Selection, Training, and Role of Technology Transfer Leaders," Chapter 4). Technology transfer is, above all, a matter of communication, putting the emphasis on the transmitter of the message between sender and receiver. That is true of all teaching systems in one or more transfer systems. The leader is the hinge of the two cultural systems; it is his job to transmit a comprehensible message, that is, one adapted to the receiver's environment.

For that reason the leaders' selection, educational training, as well as involvement in the creation and formulation of the message, are of prime importance. That is why his recruitment and training must come before the programs are established. His educational aptitudes, his aptitude for mastering and formulating the message and his sensitivity to socio-cultural differences, are equally important criteria to take into account.

Where do we find such people? They will usually be found in the receiver's region, country or, professional environment, if he is already competent in the technology to be transferred. Otherwise, that competence must be looked for outside, often among technical advisors.

Careful Programming of the Teaching (see "Creation of the System," and "Training," Chapter 4). The teaching itinerary within the system must be conceived in such a way that the teaching content can be placed within an educational progression so as to acquire well-assimilated, immediately useable knowledge and know-how, and to allow for retraining which might prove necessary to meet possible failure or rejects.

And this is a fundamental point: the training's effectiveness and permanence of results are directly related to the educational quality of these itineraries—as well as the reduction of direct and indirect training costs.

Choose the Educational Processes and Prepare the Teaching (see "Training" and "Group Training," Chapter 4). The classic educational processes and methods used are: the conference or direct tuition, group lessons, individual or group work, case studies, and laboratory or workshop classes. In addition, the professional environment permits useful training procedures such as visits to plants work on professional projects, and, above all, group courses formed by the leaders.

Yet these courses can differ greatly. Of either long or short duration, they can vary from simple learning and observation courses to applied courses in a professional environment. And they can go as far as the taking over and running of professional operations by work teams.

Finally, they can call on an increasing number of audiovisual aids: television, video-tape, tape-recording, magnetic charts, films, transparencies, simulators, etc.

Without being exhaustive, this list demonstrates the need for careful preparation; each teaching situation must have a method best suited to it.

Establish a Permanent Exchange with the Socio-Professional Environment (see "The Dynamic of the Whole," Chapter 4). It is advisable to create the conditions for a symbiosis between the teaching system and the socio-professional environment by setting up systematic and permanent exchanges between them. Such exchanges can involve:

1. Studies of the profession's structure and its capacity to receive, followed by periodic questioning of the teaching system's aims.
2. Professional participation in evaluating trainees during and after training.
3. Joint organization of courses.
4. Participation teaching in various forms by people from the professional environment, and their return at regular intervals to professional life.

Apply Permanent Control (see "Measuring Results and Corrective Action," Chapter 4). This must concern the quality of teaching, its methods and means, as much as the trainee's progress.

Applied at the trainee level, the control has two aims: corrective (catching-up and orientation) and educational (self-evaluation, integration into the educational method, discussion on the results obtained).

Applied at the educational tools level—programs, classes, teaching aids—the control must be a permanent training concern; but it must be realized that good intentions are not enough; for that reason the control must be institutionalized, its frequency and supporting aids defined and formalized. Without structures or procedures, the hope for educational and technical progress will not be realized.

Organize the Teaching System (see "Organization," Chapter 4). In order to define the teaching system's structure, two dimensions must be taken into account: the implementation of educational methods and the teaching content.

On the other hand, the educational methods impose a certain functional organization on the system, whether it concerns the preparation and realization of programs and classes, or the teaching itself and its supporting activities, such as the fabrication of educational aids, the organization of on-site courses and classes, etc. At the same time, the administrative services must adapt their structure and behavior towards pupils and teachers to harmonize with the spirit of the particular educational method(s) selected.

On the other hand, the teaching content, which arises directly from the economic aspect of needing personnel, demands strong similarities between the structure of the industry and that of the teaching system.

And finally, when the system's creator gets a chance to participate in designing the training schools, he should not fail to point out the importance that environment will have on trainee's behavior; he should also insist on the need to adapt the architecture to the educational procedures.

Recruit the Trainees (see "Recruitment," Chapter 4). Above all, recruitment must be based on real motivation and actual or potential capacity for work by the candidate. While the classic tests of knowledge will reveal a trainees' starting point, they will not show what he is capable of tomorrow. By contrast, psychological aptitude tests give much more valid results—as long as they are adapted to the population's socio-cultural conditions.

In fact, experience shows that a trainee rejected by the classic teaching system is not condemned never to learn, nor totally unadaptable to working life. Practical training and its application are a powerful means of recuperating them.

Create an Overall Plan (see "The Transfer Strategy and Overall Plan," Chapter 4). At the end of this analysis of "How to Create a Teaching System," it is right to insist on one last point—the importance of planning in a teaching system project. In fact, each of the stages described above must be planned so that the people come on to the market when they are needed. Tomorrow's needs will only be satisfied by taking the right action today.

6

Genesis of Our Method

First Thoughts

The method described in Chapter 4 and its extrapolation in Chapter 5 cannot claim to be absolute. Its value lies in its effectiveness measured in action when confronted with widely differing sectorial, ethnic, and cultural situations. Moreover, it is born in action, developing through operations that aim at success, but which sometimes must face unexpected obstacles, remaining content with a more modest success, while avoiding failure at all costs.

Why repeat these steps? Because only they can bear witness to the proposed method, which, without experience, would have simply an intellectual value.

Before presenting our work, it would probably be useful to put into context the beginnings of our applied research, such research being inseparable from the consultant's profession.

Ours is one of the most recent kinds of management consultancy. Having for some time integrated and been enriched by specialists, it goes beyond the simple technology transfer consultancy to which we often refer in these pages. Its aim is now to aid the development of companies and administrations. All developments, in fact, result from a successful transfer (either a grafting from outside, or from inside) in the company built around one person's forward-thinking idea.

Management consultancy has found a place in all the developed nations, which now number thousands of consultancy practitioners, sometimes grouped in large companies of several hundred employees. Now that it is recognized, management consultancy can—and we say should—move into the privileged domain of applied research. But it must face certain constraints both in the approach and in the development of operational fields, before which we must continually yield.

The Features of Our Approach

A Practitioner's Approach

A good-sized consultancy carries out more than one hundred commissions a year and harvests an extraordinary amount of knowledge from management colleges which it would be unforeseeable not to use.

Today we can reflect on a thousand commissions carried through in thirty-five different countries, in about thirty industrial and government sectors. Some large technology transfers have been carried out covering several hundreds of thousands of people. The employees involved everyone from illiterate manual workers to chief executives from the top management schools. Some 300 educational manuals have been written for teachers from different management grades whom we prepared for new tasks. These documents represent around 800,000 hours of prepared teaching, having been prepared several times, each time to groups of eight to twelve people.

A Foresighted Approach

Each new operation must be a success and everything will be done to prevent its failure. In this domain professional ethics unite with concern for our image. One bad reference has a greater effect than a hundred good ones, each bad mark being a heavy handicap in our development.

In the development of a consultancy there must, therefore, be an approach that leads it towards greater complexity and diversity, and replies to the needs of a market facing the same development. But at no point in the growth curve must there occur a blunder which could prove costly for the client.

Not the least of difficulties encountered by a management consultant is one common to all professions (even surgeons), and that is, when faced with a new commission, whether previous successful experience can be used in a way acceptable to both parties (particularly the client) or whether the break with the past is so great that it cannot honestly be applied.

Our step in the direction of greater complexity and diversity is thus a foresighted one whether it concerns the size of a group, the nature of an industrial sector, the particular features of the country receiving the transfer, the management level of the men involved, the company's business, or its top management.

As far as operational diversity is concerned—production, marketing, personnel, data processing—and to some extent sectorial—oil, mining, energy, transport—our experience reveals that there can be no growth (except if we take risks that normally we would seek to avoid) without integrating one or more outside operational or sectorial specialists into the basic team.

As long as it is carefully prepared, the grafting of a new function or sector will succeed where it unites methods drawn from long experience in certain industrial functions and sectors, backed by extensive expert knowledge. The barriers dividing each person's skills can thus be crossed without conflict.

An Open Approach to Outside Influences

If he is honest, the consultant must be open to all innovation, all information that validates a method, all progress and knowledge that will aid the success of his actions.

He must be orientated, therefore, towards all universities, research centers, and even to avant-garde companies that introduce organizational or training innovations. But he must also be open to certain writings emanating from various schools of philosophy which, though not concerned with contemporary problems, often produce messages with a contemporary relevance.

This openness to external influence was particularly noticeable during our first commissions, particularly in the experiences of our small team of 1960. Our line of development thus begins from two sources.

The first is the commissioning of new plants. Since 1950 I have been participating with Electricité de France in starting-up, and training the labor forces, for 14 power stations. To that end I visited a large number of power stations in West Germany, the UK, Sweden, and the United States which had been put into service during the active 1950-1960 period.

The second source was adult vocational training. From the beginning of the first productivity missions to the United States most of us had joined "Training within Industry," TWI, programs. Some of us conceived new applications and found in the method's instructional sections an efficient guide to creating detailed training programs for a specific task, or the group of tasks that makes up a job.

Moreover, the lessons of the French and Allied was effort, and the concomitant training effort using adult educational methods had left its mark on a number of us. Several had benefitted from Carrard's (Swiss psychologist) lessons that had been applied in France before the war.

An Experimental Approach

In working with people, and especially groups, the only method that leads to real progress is the one that is successfully used in all other sciences—the experimental method. But such an affirmation should not lead one to underestimate the difficulties of a life-sized experiment in part of an economic body that is permanently confronted with the struggle for productivity and sometimes survival. Are these difficulties any greater than those confronting a doctor? And the risks involved, (limited by the extent of the expert's experience and desire for slow progress) are they any less tolerable when it is a question of a company's performance than when it is a question of a human life? We freely admit that simple or complex models can aid the doctors as well as the company executive. But when we are faced with beings as complex as men nothing can replace life-sized experimentation.

What is most important is to combine modular studies with progress in action—and above all not to omit any of the six essential steps in the experimental method:

1. Observation of phenomena
2. Selection of relevant sizes
3. Construction of a representative model
4. Extrapolation from the model's behavior
5. Experiment or proof
6. Definition of the field of validity

Detailing these six steps evokes an obvious similarity between the experimental method and our educational method proposed in the training arc in Table 4-1. Five of these steps are very similar to it, veering from the actual example to the model or diagram and returning to the real example for verification or consolidation. The experimental method's sixth step is echoed in the educational method by defining the validity of training tools: classes, simulators, video films, etc.

That is, the experimental method is present both in our approach and our research, as well as in the educational method proposed to trainees. The trainees are invited to discover and validate each step in their own field, moving forward in small steps from one stepping stone to another, discovering motivation in the game of exploration in which they are permanently involved.

Introduction to the Chronology

Two methods can be used to describe the slow maturing of the method described in Chapter 4.

The first is a question of unravelling the sequence of operations. In doing so there is the obvious risk of replacing a daily reality of hesitations and piece-meal improvisation with a well-constructed fiction.

In the second we forget hindsight and recount the true story using only actual texts. That risks a travesty of the reality in the author's favor, but in counterbalance the reader must make an attempt to synthesize many different documents.

We chose neither of them. Or rather, we chose one or the other, and in that way hope to meet the needs of the largest number of readers.

The five sections that follow (grouped in three-year periods) present the milestones in our method's development and the validation of its field of application. The lower half of Figure 6-1 shows, diagrammatically, the increase in the number of areas we now act in starting from 1960 with new plants and elementary training systems. The upper half shows the development from those methods.

The seven appendixes at the end of the book illustrate these methods particularly well. They are taken from about fifty articles that we or our associates have

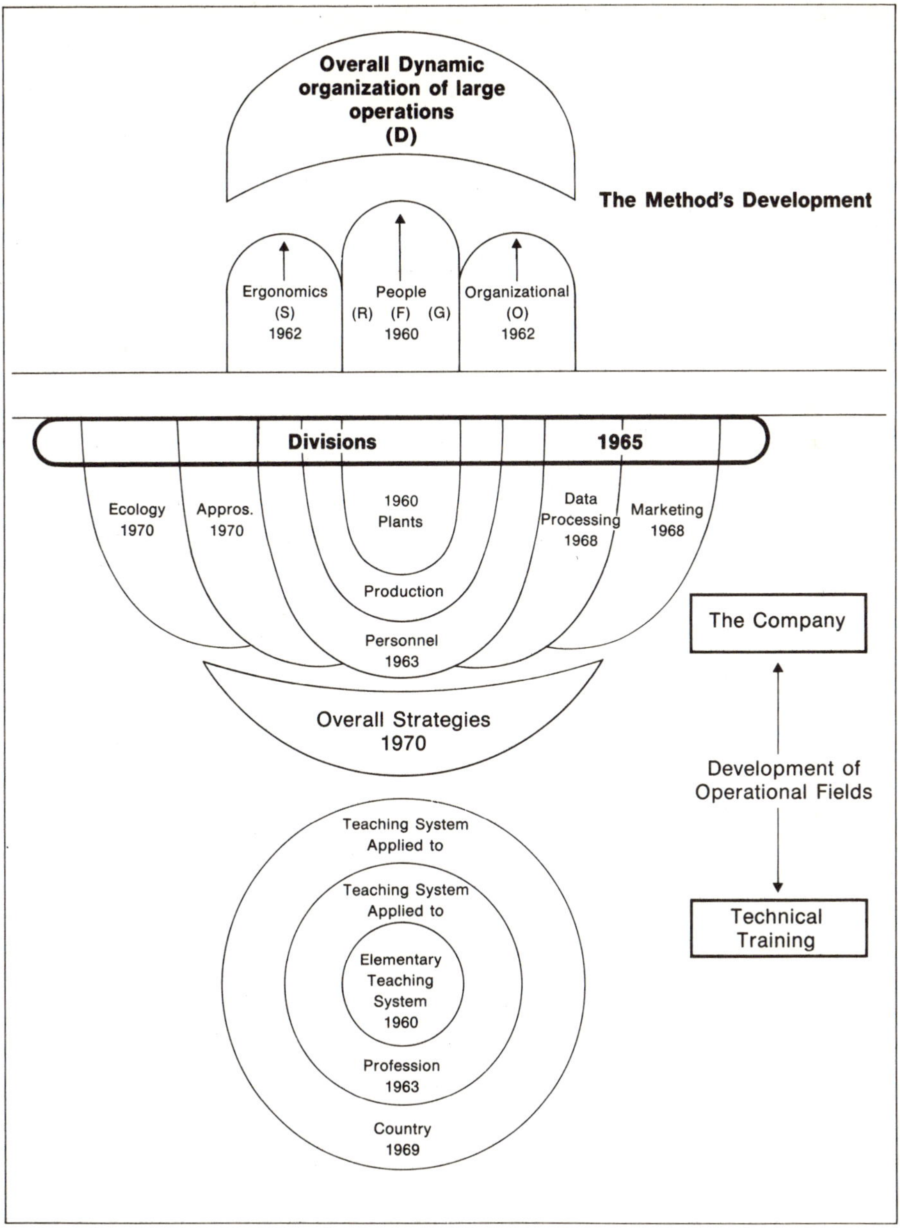

Figure 6-1. This diagram shows the development of Eurequip's method of technology transfer.

published and explore in a more comprehensive way points presented in this and the two preceding chapters.

Thus, the reader seeking an overall view of our method's birth has only to read this chapter. However, the reader looking for more detail on some points can consult the texts.

The Period 1960-63

Development of Operational Fields

From the beginning we attacked two sectors (referred to in Chapter 1); our efforts were concentrated on them for the first five years and today they still make up a large part of our field of operations.

The first sector is the preparation of a labor force for a plant under construction. This is a relatively new one for consultancies, which are traditionally more at home in an existing plant where criticism is easier. Moreover, when we began, such training was completely ignored by engineering consultancies, being more concerned with industrial processes than personnel. For us it was a matter of extending to other kinds of plants the experience gained from setting up power station crews, then extrapolating from it without heedlessly crashing very different technical or socio-cultural barriers.

We were guided by a special feature of power stations. A power station is a "continuous process" plant: that is, the transformation of raw materials—a combustible material mixed with oxygen—into a finished product—electrical energy—in a continuous operation by fixed equipment—boilers or huge turbo-alternators. Apart from procedural differences, running and maintenance in continuous process plants do not vary greatly: continuous working at fixed job points, operations directed from a central control room; shifts of machine supervisors; the need to allow for preventive forecast maintenance, etc.

So it was in continuous process plants in developed industrial nations that we began to extrapolate from our experience in preparing power station crews. The first were a large glassworks outside Madrid and the Esso Standard oil refinery at Fos-sur-Mer (see Appendix 1).

The second sector, creating a training system for workers in similar jobs, was reached more easily. The system is little more than the elementary component in a system for preparing labor forces for plants under construction. Thus, there is a difference in the complexity of the two types of operation. The plant being built presents many problems (delivery deadlines, absence of tangible jobs, the need to interact with many people), which we do not encounter when training personnel doing the same job in a company.

The sector is, therefore, simpler; our experience is much wider; our knowledge more sure; and the extrapolation curve rises more sharply as a result because it can apply to the job workers must learn as well as their socio-cultural backgrounds.

We have applied this experience mostly in Algeria and Morocco among people ranging from the illiterate to college graduates. The jobs differ enormously—from mine timberman to rural tax inspectors, from trans-Saharan lorry drivers to public comptrollers. Table A3-2 lists our operations in the Maghreb countries.

Development of Operational Methods

The framework of our operational method is already drafted in Appendix 1. In this particular case, the technology transmitter includes not only the plant operator, but also the international group to which it belonged, the consultancy, and the constructors.

Certain characteristic features of the preparation of work teams for plants under construction include training and selection procedures, such as: recruitment in successive waves, training in successive layers of managers, foremen, workmen; close liaison with the action program, and plant completion. The technology transfer leaders naturally come from the new plant's management and their role is a vital one.

Modifying the organizational structure to give personnel more incentive while complying with certain company aims (see "Organization," Chapter 4) leads to considerable changes in specific, fixed place jobs, i.e., the creation of versatile workers capable of doing several specific fixed-place jobs as well as day-work maintenance, and therefore doing less night and Sunday work than a fixed-job worker.

The training preparation that draws together management and consultants results in unique training manuals created especially for the workers and the refinery.

This huge project in a new field (the oil refinery) made us even more aware of the need for cooperation between those responsible for devising the jobs (in which we were fully involved) and the well established plant builder and technical consultant.

The project's quality can have important consequence for the men's work quality. It is the transfer architect's sad privilege to apply ergonomics in reverse when studying the worker's jobs, through an examination of the machine emplacements (actual or on paper) decided by engineers. He discovers that modifying emplacement details, or control room lay-out, could considerably simplify supervision, maintenance, or remote control tasks.

Unfortunately, when the architect discovers these holes in the installation's ergonomic chain his opinion is rarely accepted; usually he is greeted with the unanswerable reply, "It's too late to change."

We were asked by the French Institute of Petroleum to take part in training oil company new-project engineers. We drafted for them (see Appendix 2) the complex process involved in drawing up a plant project by a multidisciplinary team concerned as much with products as machinery and men. Labor force training and the technology transfer can be considerably simplified if, at the plant drawing-board stage, the people who will run and maintain it are thought about too.

From 1963 the creation of elementary training systems really began to take shape, based on the thirty or so commissions that fell into our second sector. These training systems are further discussed in Appendix 3, where we rediscover in a more elaborate form the skeleton of our method proposed in the Fos-sur-Mer case history. Note that the operational sequence in Table A3-1 is very similar to the ten-point, Chapter 4 method. The major omission is the modelling of the men and micro-structures ("Organization," Chapter 4), which, at that time, we dared not consider an integral part of the technology transfer.

This first draft of a method was carried into the field by people far-removed from a technical culture, from which it seemed possible to construct solutions to the problem raised by permanent adult education. (The French law of 1971 further institutionalized adult education).

The Period 1964-66

Development of Operational Fields

Although the massive Miferma transfer, begun in 1962, should have figured in the preceding chapter (the operation is referred to in detail in Table A3-2), we consider it more appropriate to relate it here (and in greater detail in Table A3-2 and Appendix 4), because we had to wait two years before being really able to judge such a large extrapolation of our methods.

Miferma gave us the chance to create a vast training system preparing 800 Mauritanians in the course of one year for 70 different kinds of jobs. The job's techniques differed widely and the whole operation was synchronized with the building of the plant which the workers would run and maintain.

Miferma is thus a synthesis of our two sectors and the experience gained in both of them in the preceding period. In fact, Miferma was, on the one hand, a vast plant under construction (or rather group of plants); on the other, the socio-cultural circumstances were very similar to those already encountered in the Sahara and Moroccan Atlas mountains. Miferma, furthermore, was the chance to apply a self-training organizational structure to aid the integration of Africans into an organization originally conceived for Europeans.

Development of Operational Methods

Two points of experience emerge from this period. One was that we could consolidate the three previous years by adapting the preparation of work teams for plants under construction to widely varying situations. Miferma, several oil refineries, and a petrochemical plant are examples of that. The second was that we went even further into the barely explored domain of adult education.

The work with illiterate people in the Moroccan Atlas mountains and the Sahara were unique opportunities for progress. Illiterates working in an industrial world are faced with a bewilderment that only a carefully adapted technology transfer can overcome. Their basic cultural uniformity helps experimentation and the revealing of correlations between the method and its results, which are more difficult to discern among the more heterogeneous populations from industrialized countries.

The transfer system depends very much on understanding and using the basic culture, controlling actual multidisciplinary situations, seeking synergy in the educational steps. In other words, the systematic interaction between real working situations, models, and diagrams. The arc method was first used among illiterate people and validated later on among people from many levels and cultures. That tends to show that everyone is illiterate with regard to other cultures and faced with them finds himself with a blank memory.

The Period 1967-69

Development of Operational Fields

Our desire to draw attention to certain transfer success factors led us to become interested in the interaction between the quality of the technical project and the workers' jobs. It also made us interested in a technique—ergonomics—whose object was to adapt machinery to personnel.

That in turn led us towards three connected fields, all equally vital, which we had already touched on. They now came to the fore during this period. They are the personnel function in the company, the technology transfer at management level, and an organization's self-training capacities.

That the quality of the tranfer's success is closely linked to the effectiveness of the receiving company's personnel function is a truth not admitted by everyone. And yet there exists a correlation both at the operational and strategic level.

The operational level—delays in recruitment decisions push the best candidates toward more alert companies; starting salary inequalities are a source of interminable discussion, destroying the incentive to learn; empty promises about fringe benefits such as accommodation and family expatriation, deflect trainees from acquiring skills

The strategic level—foresighted career path management, management structure modifications, promotion of native workers, termination of expatriate's contracts are complex operations. The responsibility for them lies with the receiving company as far as its own work force is concerned. It must be in complete control.

We were alert to the importance of the personnel function in a company's capacity to make a success of a transfer and development generally. For that reason during the period we faced up to the personnel function's place in the company by paying great attention to certain systems that were then still rudimentary—particularly personnel management forecasting.

Having carried out personnel management forecasting at company (Miferma) management and plant level, we intended not only to handle the problem of forecasting for increased numbers of workers, but also to provide certain statistical answers. In this context was revealed the importance of the dynamic of the whole in the method—permanent further retraining. Receiver-company management had always played an important part in previous transfers. Frequently, they were used as transfer leaders.

At the same time that experience made us aware of the "operator-innovator" duality in the manager's role. It also made us aware that in our method we put too much emphasis on the operational aspects. Of course, a manager, and particularly a production manager, must first be certain that his workers are competent at their jobs and proceed in the best way. When he is actively involved in a transfer, it is in this part of his job that a manager exerts most effort. But what do phrases like "competent at their jobs" and "proceed in the best way" mean? "Competent" has no technical meaning, and "best" is relative. A plant can always be improved, even on the verge of obsolescence.

To quote once more Saburo Okita: "On numerous occasions, thanks to similar improvement, factories built to foreign plans after awhile arrive at production levels much higher than the nominal capacity fixed by the foreign know-how supplier. Steel blast furnaces designed to produce 1500 tons a day, produce in fact 2500 tons. A synthetic rubber plant with forecast capacity of 50,000 tons a year produces 70,000 tons."

Managers' innovating role in an area where it would seem to be negligible, or even prohibited, can thus be considerable. But having said that we are not ignoring the obvious criticism which can be made: beware of the risk of permanent perfectionism by the plant operator. Yet, can it be called perfectionism if we try to draw the daily lessons from a breakdown, to prevent it from happening again; to reduce wastage by better machine tuning; to improve a measuring device's accuracy?

It is too easy to say that the company is carried forward by the momentum of progress. If it allows itself to be towed along like that it might well end up "on tow"! If, and above all, the company calls on foreign technology, there still remain many opportunities for improvement and innovation. Only company management is really competent to carry out such developments.

On the basis of such remarks, and in a straight extrapolation of our method, we began a rather original management retraining program at this time in French and African companies. In what way was it original? It was original in the way trainee groups treated real company problems. Here we come across training as described in Chapter 4, with a freedom of action that went beyond the group. It was not a question of using a redundant actual problem (which is often done in technology transfer), but of dealing with a real current problem in its actual environment. The aim was to come up with an answer and to positively develop the group of managers.

From the moment we touched on the personnel function and management's innovating role, it enriched our methods. We will not dwell on those advantages, but mention others:

1. The formulation—in a form which has varied little since—of an approach for preparing labor teams for plants under construction
2. A cost forecast of such preparations
3. The features of an organization open to innovation add technology transfer ideas
4. The impact on our educational theories of programmed training theories and techniques

Development of Operational Methods

The steps in preparing a labor force for a plant under construction, based on twenty or so operations in industrialized and developing countries, are much the same as those proposed in a 1967 article whose basic tenets are still valid today.

But the description omits an important detail—the operation's profitability. From our contracts with many managers responsible for starting up new plants we distinguish two managerial types. The first are those convinced of the need to prepare the labor force, who, while paying attention to training costs, do not consider it worthwhile to get involved in economic calculations to justify the amount spent and ascertain the speed of that investment's return. Often, those managers have been through the drama of a plant starting up with a last-minute labor force. They know that such improvisation can be costly in terms of lost production and worker and machine accidents. For them the risk can in no way be measured against certainty.

The second category (just as numerous, but becoming less so) is made up of those who think perhaps with justification, that "one becomes a blacksmith by blacksmithing." They refuse to spend large amounts on preparing the labor force. The holders of this opinion are frequently found among engineering consultants who are convinced of their jobs' difficulties. They are hardly disposed to acknowledge the amount of experience and professionalism that the plant operator must mobilize. They are also found among plant operators with a similar kind of plant in operation—but who did not live through the disconcerting start-up days, when a vast complex of personnel and machines must get under way.

So as to reassure the first in their choice (by clearly enumerating the arguments) and to unsettle the convictions of the second, we drew up the training balance sheet. Such is the aim of Appendix 5, which demonstrates that training is rather like an insurance premium and that its real cost can only be compared with expected savings of more reliable management.

This test deliberately does not take into account any wishes of the plant operators who may want to go above the economic level and for political reasons put the maximum number of native workers into the plant at start-up.

What do we mean by organizational structure's training value? Or, by extension, its development value? We had carried out earlier work on the adaptation of organizational structures to workers' capabilities (by extending the ergonomic effects of a concrete job environment to its organizational environment) and career paths that permitted self-training. That work made us well aware of the close connection between a successful transfer and the planning of the management structure.

What is true of a simple isolated job is even more true in an organization such as a plant, service department, or whole company. We have come across companies that assimilated new technologies easily and quickly, while others, apparently identical, were unable to.

By examining this favorable company behavior at job level we (the author and Dr. Marcel Filiu) formulated the following theory, validated by later work:

1. Each elementary man-job system contains two end products. One is a work output, that is, value added to a material or product for internal or external company use. The other is a basic transformation in the man; he is not quite the same after the job as he was before.
2. Organizers usually pay more attention to the first output; it is also possible to pay attention to the second so as to determine which company features develop workers and which retard them.
3. It is not a question of moral choices; whether we like it or not, companies change people.
4. The company's development capacity can grow at the level of each job. That capacity is characterized by six systems which "irrigate" each job in the company. Those systems are: planning, delegation, communication, internal incentives, external incentives, and judgment. Each job "irrigated" by these six systems develops the maturity of the job holder; he brings a more complete approach to the job while prepared for greater responsibility.

Among managers this capacity in the organization for developing workers (which we call "anthropogeny") reawakens basic elements alerting the managers to their creative capacities, positive commitment to progress, and innovation.

Let us now look at the impact of programmed teaching theories and techniques on our educational ideas.

We have been completely open to these techniques and theories ever since they were introduced into France. We have toiled long enough preparing training classes to be able to welcome any development that lets us go beyond the simple "art" (if not craft) of purely pragmatic training.

This openness is not just intellectual. We believe in action and the progress leading from action. Our team created a number of programmed works on various subjects such as financial management, computers, computer peripherals, and decision logic.

From the confrontation between the lessons we learned from working on the programming and the conclusions drawn from preparing classes to be presented by teachers evolved a synthesis between rigorous American methods originating from universities and the pragmatism of our own work, born in action and on the job. This synthesis had already been presented in "Training" of Chapter 4.

The Period 1970-72

Development of Operational Fields

Development was systematic during this period and the following one. It progressively covered all the major company functions and its specialized techniques to which the previously validated transfer method had been applied in a more limited area. Also begun were, in order, data processing (after many precautions, this was begun in 1965), marketing, and very cautiously, strategic planning, which should more naturally be placed in the 1973-1975 period.

With the benefit of hindsight, the penetration of data processing into companies and administrations can be seen as a unique opportunity to confirm our experience in a new area. Refusing to join the general jubilation of the high priests of this "art," we had from the very beginning considered computers as part of the unceasing development among companies, their objectives, techniques, and employees' jobs. Machines appear, are perfected, collect, handle, and distribute information. As in all companies, information is the raw material of decision-making; these machines and the network surrounding them will become the nerve centers of the most brilliant companies as well as the most modest, the most innovative, and the most routine; and to that extent will be present everywhere.

One of the first impacts of computers in companies will be on new plant buildings—a plant without walls, but not without reality—a plant which will mobilize a labor force composed solely of specialists, but which will spread its tentacles throughout the organization, modifying the jobs of many workers. Of course, like all new technologies this one will meet resistance to change. To become accepted it will need to pass through the stages of differentiation and integration before it becomes commonplace (see Table 3-2).

To sum up: a new kind of structure will try to take over the company. But it will not penetrate unless management is convinced of its usefulness. And that penetration is reflected in profound changes in many people's jobs, and they will not be immediately favorable to it. Let us add that there can be no mistakes; the new man-machine system will be more resistant to change than the systems it replaced; the inertia of the machine and its programs are greater than the workers'.

So it was by training and the creation of a microstructure favorable to training that we approached the opportunity of the data processing tool and methods. Management problems are now solved by computers. The research department, for example, had discovered the power of this new tool and, thanks to it, the chance to widen its search for optimal solutions. As resistance to change withers, the field open to data processing widens. But the computer must not disappoint us by creating in its track jobs that recall the worst excesses of mass production.

Our preoccupations at this time are found in a portion of the text delivered at the "Management, Automation and Humanization" Conference in 1969: "Let us follow this thought. We hope that the manager, responsible for a system, will develop it in the direction of higher efficiency which will lead him to accord a greater place to data processing. If that happens he will, on the one hand, change the character of his job: on the other, change the character of his subordinates' jobs."

Modify his job (and I did not say his goals) so that those goals can be more easily established with other managers and center on the company's own goals. Albert Camus' statement: "The end justifies the means, but what justifies the end?" sheds some lights on this work, which runs from daily objectives to the long-term aims of the community. Modify his job, by delegating routine decisions to machines; by delegating control decisions to his colleagues, and in this way he will play a bigger part in the system's permanent development. For, like humans—and perhaps even more so—machines resist change with their circuits and programs. It is rather like a die stamp defying the designer's creativity with its inertia.

It modifies his subordinates' jobs too through two arithmetical operations. One is subtraction: a certain number of tasks are given to the machine. The other is multiplication: the association of man and machine—more intelligent than man on his own—is capable of understanding more complex problems and widening the field of reasoning when preparing for action and making decisions.

But this only casts light on job modification at group level. Drawing tasks together (which cannot fit into the computer) into men's jobs can be done in two ways. One, and unfortunately the most frequent, is to gather crumbs of work together with no link between them. At the other extreme we try to create jobs controlling a whole process and in this way look after the men's interests too. So, a job can be like a heap of bricks or a house.

To build this house we dwelt on the need to train 5000 data processing architects (technology transfer leaders) who could conceive of the mass of jobs as well as the machines. Having said that, perhaps we were assuming the role of a Cassandra in observing: "But what will happen in the absence of these architects? Well, data processing will advance in the same way that suburbs flourished without town planners. This data processing advance—that act of organization—will be nothing like an informational art could be. By being either overly generous to machines, or by failing to understand their well-being, the spread of electronic data processing leaves behind manager and practitioner alike. And one day we will speak of a crisis in systems, in the same way that we talk about a crisis in architecture."

Marketing has also seen its technology transfers. That is rather to be desired in France where many companies retain in their management organization and attitudes a Ptolemaic view of the market place, centered on the company which alone knows the truth. The Copernican view of the market alerts companies to customer needs, signaling those needs to the research departments, together with technology transfer operations successfully carried out from a kernel of leaders, collecting and dealing with real problems relating to each stratum in the population.

Apart from these new fields, we naturally continue to work in plants under construction. In Chapter 2 we described the technology transfer example of our thirtieth training operation for a plant under construction: Somaïr. This case has other features than simply being the thirtieth. The culture shock between sophisticated techniques and the basic culture of the workers experienced at Miferma was even greater at Somaïr. The nomadic population was similar to Miferma's, but iron ore gave way to uranium, a mineral which must be enriched at the mine using elaborate processes.

We should also mention the systematic use of models and a life-size simulated control room (the plant did not yet exist) as well as the encouraging results obtained.

By the use of these devises we got over the potentially trouble-laden transplantation of a trainee work force to a reference plant in an industrialized country. The reference plant (similar but not identical) existed, but was in the center of France. It was certainly possible to send the plant operator there, but the shock upon the workers can easily be imagined. It is best if such courses are reserved for management which can be trained to collect information needed for training lower management grades. A similar training operation, using simulators and models in the absence of a real plant, seemed impossible for Fos-sur-Mer; the Somaïr example showed we were too cautious.

A similar case of a control room made from wood and cardboard is related in Appendix 6. On that occasion we did not come in at the construction stage: it concerned the prototype of a natural gas liquefaction plant. Management wisely decided that because of the plant's experimental nature, technical risks should not be compounded by human risks. The plan thus started up with an all-

European labor force with similar experience in oil refineries. The Europeans were replaced with Algerians several years later when their skills were satisfactorily developed. Replacement was spread over five years.

We should like to stress that self-training career paths were used to train the control instrument regulators; they had a good basic education, but no technical experience.

Development of Operational Methods

It is mainly management that continues to occupy our attention. We are now in full control of the operation which turns out an imitation company based on a reference company—as long as we have enough time and technical back-up.

But it seems that whatever the country's development stage, imitating performance is not enough. Training management for innovation remains a major preoccupation. We now pay great attention to American work on creativity and its relation with the theories of various European schools of psychiatry.

What place should we give to these creative methods which are undoubtedly a new technology? It would be a pity to reject them; they are undeniably effective. Should they be holed up in a company grotto like Greek oracles while problems are brought to them? Too many solutions of that kind spur us to reply that creativity concerns everyone with decision-making power—that is, at least all managers. The techniques, rather than letting them float around the company with no precise function, must be integrated, like any other tool, into the approach to problem-solving.

That approach must clearly marry rational thought which chooses, decides, questions, and, even with irrational thought, detects the problem or imagines different solutions. Creativity without decisions and action becomes fantasy, while rational decision-making, unrelieved by the possibilities of creative methods, turns up a poor choice of alternatives.

A further enrichment of our management training method occurred with the evolution of "problem handing" into "project operation," in a straight line from the development of training. It often happens that management is shaken up by creating horizontal project groups responsible for a particular operation. Such operations can be very varied, for example, plant construction, a product launch, a better way of welcoming young managers, improvements in working conditions, and, naturally, a technology transfer.

There are always one or more objectives to be respected—usually time and means such as money, men, subcontracting, etc. A leader in the project group who is distinct from the group's chairman means the project may benefit from lessons drawn from similar projects that have been put to the group. For the group members it implies a gradual improvement in their behavior within the group.

The Period 1973-75

Development of Operational Fields

The Algerian technology institutes provided us the chance to appreciate and spread the lessons drawn from the field of technological instruction and also to extend our teaching system's approach. The Algerian problem was happily one of objectives and not means. One part consisted of setting up an intermediate body of men between managers and manual workers (a group notably lacking in all developing countries); the other was to up-grade a vast young population that had been rejected by an educational system not always suited to the country's cultural facts of life, being based on elitist selection by examination.

We were not to graft a foreign teaching system; on the contrary, we were asked to create a new, original tool clearly fitting given objectives.

Other company functions were then considered. One dealt with a consultancy specializing in the supply function. This was an occasion for us to try out a technology transfer on ourselves in which we sought a synthesis between the consultancy's supply expertise and Eurequip's technology transfer method.

That operation took place at more or less the same time as another function appeared—the ecological study. Enrichment of a company culture is a continuous technology transfer. In that way we experienced the development of "human sciences," the development of decision-making, and more recently, the development of data processing. While the energy crisis forces companies to be preoccupied with short-term worries, survival and maintaining employment, we believe that they will ultimately have to face ecology. They should appreciate their company's ecological impact and the way it can be improved other than through the trenchant opinions of pressure groups.

But the major feature of this period is the application of our technology transfer method to overall company strategic behavior. In order to start on this work we enlarged our team with new specialists in this field. And we concluded an agreement with one of the world's leading experts, Professor H. Igor Ansoff.

What is now taking place is a three-actor game between the expert with the know-how, H. Igor Ansoff, Eurequip's new associates—chosen for their experience with huge multinational companies—and the rest of Eurequip strongly supported by its technology transfer experience, which it has applied in other areas at a more modest level.

Appendix 7 describes the results of this triangular transfer. Pierre Davous, who wrote it, has made a synthesis between our previous educational steps and his personal experience as a manager in a vast multinational company where strategic techniques and behavior were introduced. Management's retreat before strategic thinking generally reflects two gaps. The first is the lack of tools for strategic thinking. The second gap relates to the manager's attitude to the uncertain future. "We suffer the future, and think about it when it's the present..." However, the strategist's attitude should be different (considering himself partly the creator of his own future) if only because of his ability to perceive and react to it.

Two developments are therefore needed: technical and cultural...to which must be added a third, organizational. Even if strategic thinking involves management grades right up to the ultra-secret summit reserved for top executives, it must also concern many managers.

The methods described by Pierre Davous were first applied in France among twenty or so companies, then among multinational companies in Europe. Today they extend to companies and administrations in developing countries.

Development of Operational Methods

During this period our efforts were concentrated on expressing the method in a sufficiently "ecumenical" way so as to apply it to a vast range of activities. It was therefore a landmark in our progress, an awareness of what often seemed an unlikely coming together of different roads.

At the same time, having classified our work's results, we can add to our method various new specializations (determining the time a person needs to train for a job, to spend in a job before promotion, etc.), which make the method operational; the continued validation and publication of that method is our present objective.

A further focus can be brought to bear on the techniques—and perhaps one day the science—of the conscious development of human groups. Various neologisms have been created to describe this group of techniques. We use "transformation," "evolution," and "socio-dynamic." But there is little point in pushing against society's locked doors without devoting some time to finding ways of opening them.

7

From One Man to Another

Various aspects of technology transfer, legal, economic, even political, are in their turn highlighted by current events, and deserve further, merited thought. But that will clearly be of little use if the transfer does not take place. However, we must admit that technology is not transferred as easily as a material product, through a simple contract, or a contagious disease, by contact. That is why the actual methods and procedures in a transfer, rather than the aspects quoted above, are emphasized in this book.

What conclusions do we draw? We present four, the last of which takes the form of a proposal—or rather a wish.

Technology Transfer Is a Human Characteristic

The slow, then unstoppable, ascent of man towards his industrial future was only made possible by cross-fertilization—vertically and horizontally—permitted by technology transfer. It is vertical from one generation to another; horizontal from one people to another. History contains numerous lessons transposable to our times, which make current ambitions more credible. And that leads us to our second conclusion.

The Transfer Is Usually Possible

Yes, but on condition that we do not hope to achieve everything at once; that we admit that it takes time to move towards more complex operations; that we admit the need to master the running, maintenance, and setting, and then the construction of a machine before aspiring to make it; that we admit the need for pooling knowledge by bringing together the men and women who possess it, both being helped by free enterprise. While a transfer is always possible, it is counterbalanced by the need for an exact understanding of its complexity—which leads to our third conclusion.

The Transfer Is Usually Complex

The facile notion that transfers are simply a matter of carbon-copy transmissions, which is so easily gained from a touristic visit to a country, must be categorically denied. We are forced to echo Professor Herbert Giersch (director of the Institute for World Economic Studies, Kiel, 6th European Management Symposium, Davos 1976.): "Technology is not exported as easily as some countries suppose."

History, moreover, contradicts that facile notion. It abounds with examples of technology borrowing from one people being adapted to another by the action of their individual genius. And it underlines the need for individual endeavour which is industry's major need. Where that is concerned we must applaud Mohamed Brahimi when he observes: "The major problem of our time, technology transfer, demonstrates that progress is born in pain."

But despite that impression of the so-called hard life, held for different reasons by people on both sides of the transfer, the receiver sees in the carbon copy the recognition of his country's aptitude to receive the technology, while the sender sees it as simplifying his job. And yet how much more realistic, and in the end more exciting for everyone, is the transfer which takes into account all the industrial, cultural, and human realities of the project. But such a course creates confusion by its innovativeness—which leads to our fourth conclusion.

Transfer Methods Are Poorly Understood

Too many transfer actors admit to presenting a transfer operation as if it were a mysterious black box in which only what goes in and what comes out are of any interest. But technology transfer, like any body of knowledge, will not advance unless industry and science manifest a real desire for international cooperation.

The transfer will certainly involve discussion at the highest government levels. And so it should because a successful transfer is the key to real progress and that is certainly a political event. The transfer will certainly involve contractual discussion among public and private companies. And so it should be because from the receiver to the lessor and including the consultant and the other interested parties, each has an interest in seeing his expectations and commitments expressed in clear legal and financial terms.

But politics, industry, and commerce will never replace knowledge and that cannot advance outside certain channels. Everyone agrees that neither international government-level commitment, nor a legally correct contract at the highest industrial level (whether turnkey or products-in-hand) will rid parts of the world of the curse of cancer or heart disease.

The channel leads in a different direction: through experience, sometimes through success, often through failure, always through using the results of both. . . and again through experience. Let us state clearly that technology transfer professionals would gain by following the example of the medical profession, continually publishing the lessons of their successes and failures. So rejected heart transplants are widely published while more numerous transfer failures are never published. It is in hushed voices, secretly, that we talk about the too-numerous, expensive plants which, at best, deliver production greatly below that promised.

We should climb out of the atmosphere of mutual suspicion, free ourselves of legal disputes to arrive at an understanding. Each case, each success, and above all, each failure, should be communicated, revealing the facts, the solutions, and the results and trying to establish correlations. Universities and international bodes should find in those published works a witness to experience, ample food for thought, teaching and discussion. Only in this way can one or more voices be heard and technology transfer reach maturity.

Obviously, a major condition for progress in this direction is the recognition of the autonomous and individual character of this publishing contribution. That leads us to mention a danger not always seen in 'products-in-hand' contracts by which one hopes to obtain in an overall contract the tools of a technique—machines, buildings, plants—and the guarantee of its mastery by the company employees. (The product-in-hand contract is agreed with a sole supplier whose contractual obligations demand that the plant he has built delivers a production level and quality with a certain percentage of local labor.)

A satisfactory contract can only be the legal envelope for a tangible entity which both sides understand. The product-in-hand contract offers a merit that cannot be too often emphasized: the reasonable wish by the plant owner to have a plant delivering the right products at the right rate. But on the other hand it presents the inconvenience of introducing confusion between contributions which are difficult to measure. And it is to be feared as the instrument of a double disillusionment involving both client and supplier in its failure.

However legitimate are the aims of a products-in-hand contract, in its most elaborate form it confuses two kinds of contribution, corresponding to two totally different tasks. One is costly and well-known (about 90 percent of the total): the plant, the contractor, or technical engineering. The other, less well-known and less costly (about 10 percent including trainees salaries during instruction): the technology transfer which is a joint undertaking between the receiver, the sender and the consultant. Inevitably the latter contribution suffers from being included in the same contract. Its importance in people's minds is proportionate to its cost. And this cost will be even further reduced as the service appears less defined.

The same confusion is found at the management level. It might be total confusion by the supplier of a turnkey plant. However, with a transfer confusion is shared between the receiver, sender and consultant. Such an operation cannot succeed unless there is a total commitment by all the actors concerned. Heading them is the plant owner, clearly exercising his job of industrial manager which there is no question of him delegating. It is up to him to set up the transfer problems in their global aspect, in accordance to the natural developing life of every organization. When starting, a plant cannot be operational if it is not really "in hand."

And that leads us to the conclusion of our conclusion: technology transfer will not advance, will not reach maturity in a reasonable time, except under varied conditions that allow it to benefit from the free action of the experimental method on a world-wide scale:

1. Recognition of technology transfer as a unique contribution justifying special treatment.
2. Wider use of contracts that involve only the interested parties and stipulate the receiver's expectations, the references and means of the lessor and the consultant. A search for satisfactory, but realistic, terms for the client between the supply contract and the technology transfer contract.
3. Replacement of the overall guarantee idea by the more fertile notion of a continuous control over trainees' progress by the instructors.
4. Systematic publication of the results of all transfer operation preferably in a text signed by all the partners in the operation. That obligation would have the further advantage of figuring in the contract and would be a strong stimulus for all the signatories who would seek to avoid the black marks which a failure would reveal to everyone.
5. Establishment of technology transfer as an academic subject in universities in developed and developing countries. A permanent liaison between the subject, the results of transfer operations and methodology research programs.

The raw material gathered there would permit the establishment of a fertile international cooperation, which we whole-heartedly wish for. One can easily imagine the annual or biennal world congress of university, industrial and consultancy specialists, gradually aiming to draw the lessons of actual experience and to trace and beat secure paths for industrial mastery. And that ambition seems appropriate because nothing is more certain than this: that all roads go from one man to another.

Appendix I

The Crew of a Refinery Under Construction: First Achievement of Maintenance-Manufacturing All-Rounders

Original title: "Creating the Crew of the Future Esso Standard Oil Refinery at Fos-sur-Mer"

Authors: M. Brossard, refinery manager, and S. Seurat

Reference: *Etudes et Realisations,* November 1964

Summary: *The future refinery is still a project. The close and fruitful collaboration between the future owners and the consulting engineers, which began two years before start-up, allows various operating options to be taken, crews to be trained and put in place, and the successful start-up of a modern, automated plant. The following is a detailed description of the action plan and the operations which preceded start-up.*

Esso Standard is building a refinery with an initial capacity of 3 million metric tons per annum at Fos-sur-Mer, near Marseilles. The new refinery will join the two refineries Esso already operates in France, at Port-Jérome and at Ambès.

Of the many economic, technical, and human problems the opening of a new refinery involves, only those that concern the recruiting and training of the "crew" are here in question.

The aim of this project is "to create a suitably trained operating and maintenance staff before the refinery is opened." Therefore, the following must be considered:

1. Entrusting the plant, which represents a sizeable investment, to insufficiently trained workers.
2. Teaching techniques of production and maintenance too soon, thus adding unnecessarily to the cost and risking to undermine the morale and efficiency of the workers by imposing a period of inactivity.

So, the aim of training staff is really twofold:

1. To ensure that each member of the staff is fully trained in time for a full capacity start.
2. To ensure smooth functioning once production has got under way. For many may find conditions during the critical launching period different from those prevailing during the normal course of production.

The Terms of the Problem

Close interdependence exists between the various processes of production, the methods by which these processes are applied, and those whose job it is to apply these methods of converting the raw materials into a finished state in optimum technical and economic conditions. This interdependence is a major problem for those planning the pattern of training. Other difficulties include:

1. Responsibility for the upkeep of the refinery is to be divided between the permanent Esso maintenance engineers who will carry out running repairs, and specialist subcontractors, to be called in for large-scale plant overhauling.
2. The refinery is to be handed over, finished and ready to begin production, by the engineering company that is building it. Its future staff will therefore have practically no access to it beforehand.
3. The selecting and training of the crew is to take approximately two years before the refinery is ready to be opened.

Possible Alternatives

Given these conditions, Esso was faced with a number of options concerning the future staff of the refinery and the organization of their training.

Breadth of Training

In most existing refineries, there is a clearcut division between those who operate the plant and those who maintain it. Thus, there are control operators for the units, specialists to set and service the control instruments and the regulators, and electricians and mechanics to service the equipment. Because of the extent of its automation, however, the refinery at Fos will normally require only a small number of operators.

These two circumstances can be reconciled by staffing the refinery with people who will have been trained in a very broad range of techniques. Thus one person will know how to supervise the operation of plant *and* how to service and set the control instruments and the electrical installations.

Having accepted the principle of a broad training for shift workers, it was decided that, at Fos, they would be divided into six identical teams, of which four would operate the units and two would service them. Each team would change its role regularly following a definite system of rotation.

This particular solution has many advantages for the factory and for the staff:

1. Each operator will be capable of servicing the units, and this must lead to his taking more notice of instructions and being more wary of damage.
2. When the refinery is closed for overhauling (which is planned to take place periodically) the operators will be able to make a considerable contribution to the work.
3. When medical reasons or inexperience of shift work make it necessary to swap over operators and maintenance engineers, the exchange can be made very easily. Operators who cannot adapt themselves to shift work can, moreover, be easily integrated.
4. Each shift worker will work fewer nights and Sundays than under the normal pattern of four- or five-shift teams.
5. The traditional antagonism between operators and maintenance engineers will tend to disappear.

However, one should note that there will continue to be a permanent maintenance staff that will not work shifts.

Sources of Labor

The idea of a multifunctional staff having been accepted, it seemed desirable to recruit workers who were already capable of performing one of the basic functions that would be required: operators, instrument controllers and electricians. The staff was therefore recruited from the two other Esso refineries in France and from local instrument controllers and electricians.

Recruiting in the region of Fos, while helping workers to broaden their knowledge of their work, also served as an outlet for local labor.

Principles of Staff Training

Having decided the general outlines of where the staff was to be drawn from and of the kind of training they would receive, Esso still had to determine the specific terms of recruitment, selection, and training. It is worth noting that these specific issues were faced two years before the refinery was ready and while it was still only on the drawing-board.

Nevertheless, Esso decided to use these two years for as concentrated a training program as possible, so that the staff would be prepared for and adjusted to their future functions as thoroughly as possible *before* the refinery opened.

Finally, it remained to be decided who would organize the training program. Could Esso shoulder this piece of human engineering itself when everyone at Fos would be very much involved with the new plant and would be following its construction stage by stage? Or, on the other hand, should full responsibility be given to a firm of outside experts specializing in such problems, but likely to be less fully conscious of the object to be achieved and of the particular contingencies of Esso? The board of the Fos refinery compromised between these two extremes in what seems to have been the most effective manner possible.

Considering the individuality of the problems to be faced at Fos, Esso hoped to gain much from the help of outside training specialists, especially as they would tackle these problems with fresh minds. Thus, a close collaboration between these outside specialists and the personnel division of Esso appeared an absolute necessity; and employees at Fos at all levels had to play as great a part as possible in the training of their future cohorts. It was on this basis that the recruitment and training program was drawn up.

Recruitment and Training[12]

Selection

After it had been decided where the staff was to be recruited, the criteria for selection had to be determined. The outcome was that the employee must be able to:

1. Adapt to working in a refinery. This requires a certain sort of character.
2. Adapt to shift work. Some find it easy to submit themselves to a discontinuous regime; others (particularly for medical reasons) do not.
3. Successfully undergo a long and intensive training program, lasting 14 months without a break. This requires intelligence, motivation, and an ability to absorb knowledge.
4. Take up a multifunctional job operating and servicing units.

[12] For simplicity's sake, this section only deals with operating staff. Of course, Esso had to train chemists, accountants, mechanics, crane-drivers, store keepers, etc. However vital the training of these workers is, it must take second place to that of the actual operating staff.

The selected operators would have to perform their jobs in three fields: manufacturing, instruments, and electricity. The selection process consisted of a series of tests designed jointly by a psychologist from the outside consultancy and the managerial staff concerned at Fos.

These tests were partly psychotechnical, to determine intelligence, character suitability, and motivation, and partly technical tests to discover what should be the starting level of the future employees.

As a rough indication, there were about five hundred applicants, of whom one hundred and eighty were allowed to take the exams, and forty were finally accepted as operators.

Training Methods

The general outline of staff training, and even to a certain extent its syllabus, is linked to those training methods that enable the staff to quickly reach the level required at the opening of the plant.

Only what are called "active" training methods can succeed in producing the required result. Teaching takes place in small groups of twelve or less.

Theoretical knowledge and its practical application are closely linked, so that the immediate concrete expression of newly-learned theories clarifies and underlines what has been taught. The progression of the training is carefully controlled; a new subject is not launched before one has checked that the preceding one has been fully absorbed.

Applying these methods in practice requires adequate material aids: specially designed classrooms and workshops, an extensive use of models, plans, photos, films, etc.

Finally, these training methods require instructors who are themselves trained in teaching and in the technicalities of their subject. The technical side of the instructor's job means that the future foremen of the plant can teach the workers, who will be under their command, about operating instruments and electricity.

The fact that this will do much to create a team spirit even before the refinery has been opened must be of great benefit to internal relations once operations have begun, which will be reinforced by the respect the foreman will already have acquired as instructor.

By the same token, the training of the instructors is guaranteed by their superiors, the executives of the refinery.

All forms of training undertaken on this pattern must have two distinct stages:

1. *Preparation:* deciding the training program, preparing each training session, construction of teaching aids. This is the work of the program technicians and teaching planners.
2. *Teaching and testing:* this is the work of the instructors, who it will be remembered, are part of the hierarchy of the future refinery.

These methods and this form of organization are acknowledged to be the most certain to make the staff of the refinery at Fos most aware of the responsibilities they will have to shoulder.

Training Program

Having seen the technical background to the training program and the decisions taken by Esso about staffing and the breadth of training required, we can now sketch the outlines of the training program providing for the different stages up to the day the plant begins production.

This program will contain, by virtue of what has already been said, the following principal stages, in chronological order:

1. The engineers and managerial staff at Fos perfect their technical knowledge and are introduced to the methods and techniques of staff training.
2. They then teach this to the foremen.
3. The foremen, the engineers, and the managerial staff prepare the training of the operating staff.
4. The foremen then train the operating staff, supervised by engineers and managerial staff.
5. Finally, before this training, some selection has to take place.

This process of selection, recruitment, and training lasted from September 1963 to July 1965, when the refinery began production. This training program is illustrated in Figure A1-1. It is worth noting that this program was, of course, planned backwards, i.e. from July 1965 to September 1963; only by going about it this way could one ensure that every stage be fully completed in time for the opening of the refinery.

The amount of time to be given to different stages of the program could, however, be gauged only approximately, considering the scale of the project, the availability of necessary material aids and of the staff of instructors, and, of course, the necessity of preparing the training syllabus. As the plan was put into practice, the time schedule had to be stretched in places, but on the whole it was adhered to very strictly.

The Aims of the Training Program

The syllabus of training depends essentially on

1. The level at which the employee starts, defined at recruitment stage,
2. The level to which he must be taught, i.e. the requirements of the job for which he is destined.
3. The time available for training.

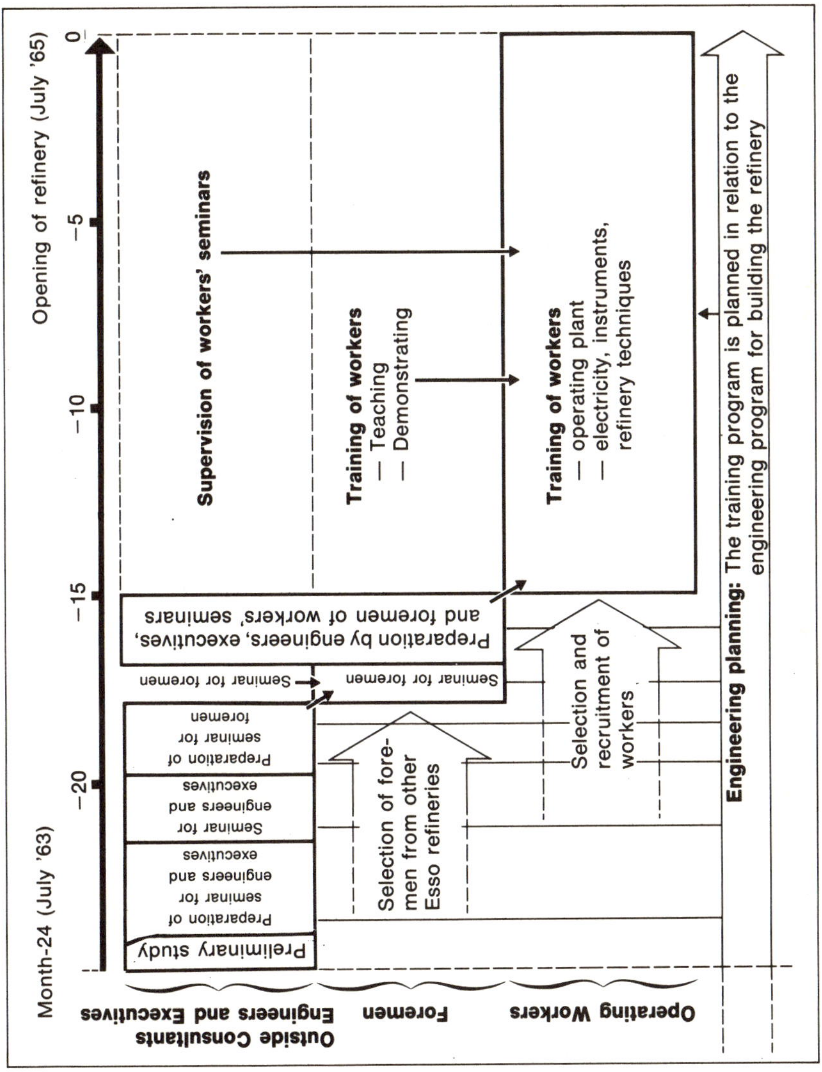

Figure A1-1. Simplified plan of a training program.

Knowing the starting level of the personnel to be trained, one must define the various jobs precisely in order to work out a training syllabus. It is from studying the job, what the individual will have to do, that one can work out what he will have to know in order to do it. Two major objectives are laid down:

1. To train the staff to perform a limited number of tasks correctly and safely, so that when the refinery opens the foremen will know the limits of their staffs' capabilities and the extent of their flexibility.
2. To serve as a springboard after production has begun to fully adapt to normal conditions of work, taking into account the eventual evolution of working posts and the staff promotion policy.

Training the Executives and Engineers

Following the advice of the outside consultants, the executives and engineers of Esso had to train the foremen and play a large part in preparing the training program of the ordinary workers. It was decided that, for this to be possible, these managers would have to be given some technical and teaching training. To this end the Esso training administrators and the outside consultants held a six-week seminar.

Training the Foremen

Since the training of the operators was to be conducted by their foremen, the foremen had to instructed, before April 1964, not only in teaching but also in the intended new techniques at Fos. Esso engineers, prepared for this at the previous stage, helped by the consultants, conducted a four-week seminar for this purpose in March 1964.

It had been planned that the foremen at Fos would have already had considerable experience of their profession. So it was only necessary to explain to them what the new refinery was going to be, what their jobs in it would be, and to introduce them to active training techniques.

Training the Operators

The training of operators had to be undertaken in three different spheres: plant operation, knowledge of the control instruments, and electricity. Planning this training involved taking into account the following considerations:

1. For the operators the refinery did not exist before the day on which production started (it was handed over by the constructing engineers with everything set for production to begin). This was a particular handicap to operation training, which had to be given at a different place and with different plant from that at Fos. It would, however, have been unrealistic

to have thought that theoretical training, however intense, would suffice. Any sort of contact with the real world of the refinery was vital. What we considered utopian in 1962 was achieved in 1965—the Somaïr case, which was our 30th plant under construction.

2. Since the prime object of the training program was to open the refinery as efficiently as possible with optimum safety for men and materials, the training in operating the units seemed to be its most important aspect. For it would, in fact, always have been possible, in the last resort, to call in outside firms to deal with the instruments and electrical matters.
3. Instead of covering all the ground of one aspect of the training before starting on another, it was thought best to alternate the different aspects in order to give trainees a balanced approach and in order not to let them mentally disassociate the various spheres, which one hoped to see integrated in their subsequent jobs, from one another during their training period.
4. No verbal explanation can give a precise description of what it is to work in shifts. The future operators had therefore to be put into shifts at the refineries as soon as possible. However, this could not be done before the trainees had acquired a basic knowledge of refining techniques.

The Assistance of Outside Consultants

Two years before the date foreseen for the opening of the refinery, its board of directors consulted a firm of specialists in the selection of personnel and professional training about the problems of training staff. Little was then known about how the future refinery was to be run beyond the principal characteristics that were outlined at the beginning of this article.

Before planning the training program itself, it was necessary to define the policies to be applied in the different areas of the plant's activity, particularly in respect of production and maintenance, where the most important problems (both qualitatively and quantitatively) existed.

This piece of work, accompanied logically by a critique of the organizational structure and a preliminary work study, initiated the cooperation between the administrators of Fos and the outside consultants. These specialists then suggested to Esso what they considered the most effective ways of facing the problems of selection and staff training. This first "preliminary study" stage of the consultants' work led to a general staff recruitment and training plan which was submitted to the board at Fos.

Once Esso had qualified and accepted this training plan, the consultants advised and participated in two neighboring spheres: one devoted to preparing the whole sequence of training, from seminars for the engineers to the training of workers; the other devoted to recruiting and selecting staff.

The part played in training by the specialist consultants consisted of preparing the seminar for engineers and executives at Fos in collaboration with three

Esso executives specially assigned to this work; administering the part of this seminar dealing with the problems of training; helping the Fos engineers with preparing the various aspects of training (work study, program planning, preparation of theoretical and practical instruction, devising of material means: classrooms and teaching aids); helping the Fos engineers with training of foremen-instructors; supervising the progress of the various stages of training; and adapting the original program in the face of incidental changes.

The part played by the consultant psychologist, in collaboration with the head of the personnel division of the refinery at Fos, involved determining a "psychological profile" for recruitment; defining the criteria of selection; preparing psycho-technical tests; training an Esso employee to administer psycho-technical tests; participation in selection exams; advising on final choices between applicants.

It may be of interest to note the length of time of the cooperation of the outside consultants: two specialist consultants during the first six months of collaboration (following the preliminary study) from September 1963 to February 1964; then a single specialist consultant for two periods of two months each; a psychologist for a total of two and a half months, divided over a much larger period.

Conclusion

In the course of their seminar, the engineers and executives of Fos studying the same problems together became a team in the fullest sense of the word, which their further collaboration only served to make yet more integrated. The same is true of the foremen who took the greatest interest in their task of training the operators.

We found that all our expectations of the practical value of applied training methods, which were used during the courses at the Port Jerome and Ambes refineries, were and are still happily confirmed. The credit for this achievement must go to the close cooperation, in preparing and administering the training program, between the Esso executives and the firm of consultants, Eurequip.

Esso's initiative in this matter lead one to attach at least as much importance to the selecting and training of personnel as to the preparation of their equipment, the classic domain of engineering. This particular case shows that the more technically advanced the machines to be operated, the greater must be the technical qualifications of their personnel. Yet at the same time, it means it is harder to find specialists (particularly plant operators) who are able to operate a plant as individual as Fos under the stated conditions.

One learns too that the specific problems of staff training are by no means reserved for plants in developing countries. They are equally crucial for technically advanced firms in industrialized countries, despite their reputation for a general availability of competent labor. Consequently, it seems that a new aspect should be added to the classic notion of engineering: the selection and training of the crew of new plant.

This task can be fulfilled by foremen (which requires, as is not always the case, the permanent availability of the necessary engineers, methods, and means) or in collaboration with specialists (as was the case at Fos-sur-Mer). Regardless, it is quite indispensable for minimizing the risks at the critical period when the plant first opens, for reaching a normal level of production as rapidly and as efficiently as possible and, finally, for creating integrated teams of workers at all levels which alone permit the profitable use of plant which is ever more complex, more costly and more powerful.

Appendix II

Impact of the Project's Quality on the Men's Jobs and on the Ease of the Technology Transfer

Original title: "Towards Multi-Disciplinary Engineering"

Author: Silvère Seurat

Reference: *Mouvements 1*, May 1967; *Etudes and Realizations*, July 1967

Summary: *Engineering is a complex job: from the conceptual beginnings it demands the participation of teams of specialists. As each actor knows his part in a play, so is the role of each member of a multidisciplinary team defined for each step in the birth of an undertaking. The work conditions of the future workers are prepared on the drawing board. This appendix also contains advice on ergonomic planning.*

Presentation

The word and function "engineering" began to be introduced on the continent of Europe a little after the end of the Second World War. Before that certain rare precursors had carried out engineering—whether they were aware of it or not—but their efforts received little attention between 1918 and 1939.

Twenty years also separate 1945 from 1965, as they separated 1918 from 1939. What better example of the relative value of time! Our age is one of forward movement, of profound changes which outdate and kill off products, plants, buildings, and entire neighborhoods to make way for new births and bring to the world those things which Lecomte De Noüy, the twentieth century French biologist and disciple of the philosopher Alexis Carrel, wrote about when referring to life: "Where evolution is concerned, nature's principal invention is death." And by that is understood: "Death which makes way for new life."

Engineering reached its peak in France under the name of *ingénierie,* which in English would be known as engineering consulting. It covers a wide range: plant engineering, ports, towns, roads, the nuclear power station in Chinon, the Miferma plant, the liner *France*, Orly airport, etc.

In France a planning commission grew up around this new profession, reference directories were published, a professional organization created; there were, in fact, so many witnesses to success that engineers began to think they had finally arrived.

And yet, no impression could be more deadly for the progress of French engineering than that feeling of success. Success, whether it be in meeting a deadline, maintaining a price, or mastering a technique, is rarely complete. It is simply the result of several compromises; there are often more unclear zones than clear ones; it is more often the preventing of failure than the affirmation of success.

In fact, nothing is more difficult than starting from scratch. And the person who complains of numerous constraints in his work is often the most hesitant and timid before the blank sheet that offers him the greatest freedom to create.

Engineering's complexity is demonstrated by certain unfortunate experiences. Depending on the case, certain partial failures (which the optimist might call partial successes) are blamed out of hand on individuals or professions, accompanied by the same complaints of "Oh, architects; Oh, engineers; Oh, economists," and so on.

Then the idea was born (and prospered almost too quickly) that engineering nearly always transcends individual specialists' activities, involving all of them in a multidisciplinary team. It is this idea of the team that is presented by certain urban planners in the context of town planning. They see town planning as involving not only architects and engineers, but also economists, geographers, sociologists, and financial experts.

Taking up that idea we shall use it for a multidisciplinary study on an industrial plant—say a steel or chemical plant, or a power station even.

Such a subject has already been widely dealt with and does not promise much. So a useful way of approaching it (and one which is more interesting for the reader) would be to draw up a diagram defining the role of various experts as the action unfolds. Yet such words as role, people, and action direct us towards an even more descriptive analogy—the theatre. So why not imagine that our team are actors in a play?

The order of appearance would be as follows:

—Economist
—Financier

—Process Engineer
—Project Engineer
—Architect
—Foundations and building specialist
—Sociologist
—Training specialist
—Plant operator
—Project planner
—Organizational specialist
—Operational researcher
—Ergonomist
—Cybernetic specialist

And the curtain rises

Prologue

The whole cast is present, the star part going to the economist. His major concern is with existing and future markets for the plant's products, as well as customers' size and location. He gathers together information on existing and future costs of supplies, transport, and fluids. He then turns towards other actors, taking from each one of them an element in his synthesis of the situation:

1. From the financier, the money available and its cost, that is, the interest rate.
2. From the process engineer, cost curves for various technical solutions according to quantities produced.
3. From the project engineer and the architect he receives value judgments on available land.
4. From the foundations building specialist, estimates for different kinds of plants.

At this stage two actors enter who are frequently forgotten. But they are not major figures in the cast. They are the sociologist and the training specialist. They are needed to estimate, together with the plant operator, the labor supply and jobs in the plant, as well as advise on procedures and automation. From that, will be estimated the annual salary and training costs. In developing countries in particular the cost to knowledge curve is not regular. Alternatives exist between a local worker with few skills and low cost and an imported worker with greater skills and higher cost.

However, it is not easy to define the jobs that can be given to workers simply because they are found near the plant. It cannot be done without weighing the local population's characteristics, skills, aptitudes, and motivation, against the skills needed for the jobs. Nor can it be done without defining the specific training to be given, and calculating its probable cost.

In a developing country, profitability can swing in one direction or another depending on the amount of high cost imported labor and the rate at which native workers will be employed. A similar kind of labor study is necessary in an industrialized country, only in this case it will concern accommodation costs, and perhaps also social and cultural centers, transportation, as well as sociological developments within the environment to be created or modified.

A new actor soon appears on whom we must keep a close watch. He is the project planner, who, from the prologue, will sketch the plant construction schedules and deadlines, refining and updating his schedule as decisions are taken.

The data thus gathered by the economist (expressed as far as possible in capital sums) should be treated as a model from which a probable result can be drawn. The economist, calculating costs where possible, planning for contingencies, and trying not to underestimate anything, will present one or more courses of action to the chief executive. His decision will cover the following points:

1. A selected site
2. A specific plant
3. The capability of producing within a specified time
4. A specific quantity
5. Products
6. Raw materials
7. A specified cost price

Before bowing out of his starring role (though he may well remain on stage) the economist will take care to construct several cost models based on the hypotheses raised in the prologue—that is, interest rates, duration, component costs, and so on. Those models will frequently be used by the various actors in their in-depth studies, and where choices must be made between variants. By juggling the economic calculations, we should easily be able to demonstrate the accuracy of our thesis and its alternative, so long as they are put in a precise framework and particularly during the study period.

Act 1

The new star is the process engineer who has already been seen in the prologue's economic study; he now steps into the limelight.

His first job is a study on the principal production line, ensuring homogeneity between the line's equipment at its various stages. He will also ensure that any machinery breakdowns will have only limited consequences by preparing contingency intermediary stocks and replacement equipment.

The principal production line is like a vast river, its flow regulated by dams (intermediary stocks) and canals (replacement equipment); its source is a lake of raw materials as it runs between added-value banks and reaches the estuary of finished products.

Tributaries flowing from subsidiary production lines or subcontractors, join the river at precise points, carrying with it parts or supplies provided by subsidiary lines or subcontractors. An example is electric motors in a car plant, or oxygen for a blast furnace.

Other networks bring in the utilities: electricity, steam, various qualities of water, gas, compressed air, oil, nitrogen, etc. These circuits originate in autonomous production centers or external sources. The network is like a series of roads, and paths going from neighboring villages to the banks of a river and its tributaries.

But to return to concrete facts, our process engineer or engineers, have prepared single-line circulation diagrams for the raw materials, fluids, etc., marked at certain interconnecting points by machinery (sometimes backed by contingency replacements) and intermediary stocks.

Without always realizing it, the process engineer is taking important decisions about the life of the plant and its operation. Maintenance policy, for example, will not be the same for a machine at a production bottleneck, as for one with a replacement or safety stocks. Nor will spare part stocking policy be the same for the example cited above as it will for differing degrees of machine standardization. Standardization will mostly affect such accessory equipment as electric motors, pumps, valves, etc.

Thinking about maintenance, about the cost of production stoppages, and downtime while awaiting spare parts is the job of the plant user, who will want to benefit from operational research. Do not forget that the bottleneck was not created by accident—it was planned as a bottleneck.

So the first act closes with a view of one-way diagrams and lists of the machines and their features.

Act II

This act opens on the laying down of principal, secondary, and auxiliary production lines. This is the art of the project leader, the machine installer, and the architect. But art or no, it does not preclude the preliminary study, nor operational research, that is the science of preparing decisions.

The major goal, besides a smooth running plant, is to minimize the costs (weighed against the total) of each transmission or handling circuit. It costs money to pipe in water, electricity, coal, minerals, and workers. It is costly to construct pipelines, cables, conveyor belts, pavements. Operational costs rise too: pumping and load losses, drops in electricity supply, loss of time, and, above all, of workers' interest.

Costs can be minimized by appropriate research carried out intuitively, or by flow charts for each circuit, the graph curve being proportional to output weighed throughout against transport costs. An example would be flow charts for a housewive's movements in her kitchen. The circuits' cost can be calculated by superimposing one flow chart on another and feeding them into a computerized operational research model.

Where it is a question of overall grouping and siting, the organizational specialist (used to studying circuits) will have much to offer, "to organize," to create the means that allows things to live. He will ensure, for example, that oil stocks are logically situated in a functional order according to equipment positioning: all equipment needing oil, everything that can overheat or run dry being grouped at the plant center with oil stocks spread around them. That is another way of saying that we have constructed the fluids and the mens' flow charts by widening the raw materials' flow charts.

Yet, if the organizer has his say on equipment location, the ergonomist has even more to say about the detailed equipment positioning. Constructing a plant on a human scale demands that we consider the importance of safety and inspection, running, and maintenance.

A valve is more easily opened, a dial more easily read, if placed at the right height. A pump is more easily dismantled if the pipes are conveniently laid out, a blocked conveyor is more easily cleared if not situated in a cellar. All that is obvious enough and the human scale is easily determined without calling on the ergonomist. Ergonomics has been practiced without knowing it. Yet, the simplest solutions do not necessarily arise from the simplest studies: in this context we should recall Taylor's long studies on the optimum size for a shovel!

The organizational chart starts on the drawing board where the difficulty inherent in certain jobs comes to light, as well as their operational safety aspects.

Act III

This act ushers in the machinery running and control network. This network is the plant's nerve center linking machines with decision centers, whether they be automatic regulators or human brains. Flow charts must also be drawn up here too: in this case they are the flow charts that carry our society's raw material—information.

To construct this network we must pick out the jobs to be filled by automated machines and the associated data, and the jobs to be filled by workers and the direct or indirect information they will need to follow direct or indirect orders.

The automated machines' field of operations cannot be defined without, at the same time, establishing the workers' jobs in order to cover the whole range of operational tasks. The workers will, therefore, on the one hand carry out jobs not open to automated machines, and on the other, fill the new jobs supervising the machines' operation and occasionally substituting for them.

Seeking the optimum level of automation demands an understanding of workers' aptitudes and their cost. The psychologist and training officer will enter the scene here, together with the automation specialist, cyberneticist, and the process engineer.

In modern industrial processes a control network starts with the setting up of a number of central operating points, each one controlling all, or part, of a production line and drawing together both the automated regulators and the information and remote control receivers. The whole—sometimes grouped around one or more electronic calculators—is controlled by one or more control panel operators. Establishing a good relationship between the control panel operator and his job, measuring the operator against the amount of information he is capable of receiving and transmitting, as well as the layout of data equipment around him, is a task calling on many of the specialists in the multidisciplinary team: the process engineer, the project planner, the plant operator, the ergonomist, cyberneticist, and the architect.

The third act draws to a close around the on-site production units, machines in place, operating methods defined, handling circuits laid down, control points and networks established. Already a large number of maintenance and operating jobs are appearing, some of which involve those who will occupy the central operating jobs.

Act IV

In this act jobs will be scheduled according to their physical location in different premises, in departmental offices, watchmen's lodges, etc. Also defined will be the information, command, reporting and consulting network linking the above positions; in other words, the final organizational structure that will materialize from close cooperation between architect, plant operator, organizer and sociologist.

The siting of peripheral buildings will now be decided too: that is, workshops, stores, laboratories etc., as well as the administrative and management buildings.

We shall start with the maintenance shops among these peripheral buildings: workshops and stores. Where should they be located and how big should they be? The answer depends on their degree of self-sufficiency as well as certain organizational options.

For example, do we want to put electricians and mechanics together, employ manufacturing and maintenance workers, manage spare parts through central administration or by the maintenance managers, place the laboratories close to production lines, and hand maintenance of automated machinery to company employees responsible for it?

We can see that the search for an optimum solution takes us through the meticulous drawing up of flow charts setting out the information paths between workers and the charts detailing human communication. If we want two workers to communicate, we should help it by placing them side by side. Failure to do so will one day demand the services of the industrial relations expert. There is a close relation, then, between communication lines on the organizational chart revealing information flows, and the physical placement of jobs as materially represented in concrete, brick, a chair, or a control panel.

Finale

It seems, therefore, that it is impossible to separate the consulting engineer's study from a preliminary reflection period on plant running methods and workers' jobs. The final report demands not so much the participation of organization experts, psychologists, training officers, ergonomists, and other specialists, as much as the use of their skills.

So, let us suppose that the curtain is about the fall on such a project—combining manufacturing, management organization, information flows, all of which are aspects of a properly functioning plant.

Final Thoughts after the Last Curtain

The preliminary project team that troops off stage is a large one. We should like to add two final thoughts about it.

The first deals with the incredible lack of first-hand experience of how multidisciplinary teams work. Of course, there is in lack of meetings between rarified and unrelated specialists these days. Frequently, they are one-day, or even one-hour meetings, with just enough time to express a point of view, a system of values, a vocabulary, without the least sign of any link between them.

That is clearly not a multidisciplinary team. A real team demands years of close cooperation around a common goal, of reciprocal shoulder-rubbing so that the rigid professional straight-jackets loosen and the team finally becomes more than the sum of its parts. A multidisciplinary team, therefore, must exist as a permanent body, meeting either in an engineering consultancy, a group of associated companies, or a large industrial undertaking.

The second thought concerns the need to coordinate the team's work. Who will coordinate it? Who will be the discussion moderator? Who knows how to get people talking at the right time? Who will know when to make the final decision when the need arises?

Replying to a similar question about the building industry where the architect and engineer are frequently alone, then French Minister of the Interior M. Edgard Pisani made the following comparison. The problem, he noted, is rather like that between politicians and administrators. The politician, who can appreciate the imponderable (what cannot be expressed in figures) coordinates and decides between various solutions proposed by the administrator after meticulous study.

The person who coordinates and manages the multidisciplinary team, therefore, will be the most "political" of them. He will be open to all opinions, understand what each is saying, and weigh various opinions against standards not found in any statistical table and which may well vary during the studies.

Let us agree with the profile of the coordinator's aptitudes. And in doing so acknowledge that the person with this difficult task (in the few teams that exist he is often an engineer, sometimes an architect) must demonstrate unflagging curiosity so as to widen his range of academic learning. No one today would dream of questioning the rigid academic divisions in higher education, but too few colleges—whether turning out engineers, economists, or organizers—are producing students with the wide range of learning necessary for the task. And yet the richness of such an education which would prepare men and women aged between 35 and 40, with, say, ten years professional experience in multidisciplinary groups to lead such teams, can easily be imagined. They would be a reply to this era's demands for exhaustive studies.

The following is a summary of the "play."

Prologue

Themes	*Actors*
Market study for the product to be made.	Economist
Siting of customers.	Economist
Money and its cost.	Financier
Possible manufacturing processes and their cost.	Process engineer
Transport and its costs.	Economist
Supply and its costs.	Economist
Land and its costs.	Architect-project leader
Buildings and their costs.	Buildings specialist
Workers and their costs.	Sociologist/training specialist, plant operator
Construction period.	Organizational specialist

End of the Prologue

The decision to build a plant at a specific spot, capable of delivering a specific product using specified raw materials.	Chief executive
Creation of mathematical models permitting common economic costings during further studies.	Economist

Act I

Themes	*Actors*
Drawing up of production lines: principal, secondary, and auxiliary.	Process engineers
Features of the machines along the lines: margins for underestimates, bottlenecks, replacement equipment, intermediary stocks, standardization (both inside the works and its associated works).	Plant operator Organizer Operational researcher

Results

Direct: Manufacturing schedules set up.

Indirect: Considerable impact on the options taken on future maintenance policy and jobs, on spare parts policy and stocks.

Act II

Themes	*Actors*
Placing of machines.	Project engineer
Placing of various manufacturing lines.	Project engineer
Drawing up flow charts to find the most efficient solutions for	
—raw materials	Organizer
—fluids	
—workers	
—optimum solution, approximate or exact, of handling costs.	Operational researcher
Placing and layout of equipment: on a human scale to aid maintenance and work, and reduce accident risks; job types and safety are born on the drawing board.	Ergonomist Plan operator

Act III

Themes	*Actors*
Creation of equipment control network.	Process engineers Cyberneticist

Flow charts and information between workers, machines, and automatic regulators.	
Direct running jobs given to the workers.	Psychologist
Running jobs given to the regulators.	Training specialist
Regulators supervised by the workers.	Economist
Grouping of control instruments in central control points.	Cyberneticist
Detailed control jobs on a human scale.	Ergonomist

Act IV

Themes	*Actors*
Placing and size of peripheral and management buildings: workshops, stores, laboratories etc.	Architect
Creation of jobs foreseen in the organigram.	
Search for the most efficient solution, using flow charts and detailing information and communication lines between workers.	Organizer
Consistency desirable between links point in the organigram, real information circuits and job positioning.	Psychologist

Finale

(by the whole company)

The curtain falls on a plant project, an information circulation project, and a jobs organizational structure project. All aspects of the single phenomenon: an operating plant which we have just created and hope soon to see come to life.

Appendix III

Lessons Drawn from Various Transfers: First Drafts of a Method

Original Title: "Training Men, the Key to Development"

Author: Silvère Seurat

Reference: *Revue des Ingenieurs,* April 1963

Summary: *There are no ready-made solutions to the training problems in developing countries: cultures different from ours, many different jobs to be filled, insufficient qualified teachers. An industrial solution grows from experience; it essentially concerns:*

- *—the rigorous preparation of properly adapted training*
- *—the training of instructors from the receiver's environment*
- *—the reorganization of tasks in simple, easily-understood jobs which is training's aim*
- *—instruction of small groups of the same level*

This appendix contains some examples of technical training for illiterate workers motivated to learn increasingly complex job situations.

Having observed that by availing themselves of scientific and technological progress, developing countries can rapidly improve their situation. The French economists J. Fourastié and C. Vimont, in their "Histoire de Demain," go on to say that "this progress has no chance of being applied rapidly on a large scale in these countries," for, the authors stress, "Production cannot be improved without the work of the masses and the work of the masses cannot be improved without first training men to be able to understand and apply modern techniques."

This sums up in a few words the real problem of the economic take-off of new countries. Specialists in economics, organization, geology, hydraulics, agriculture, and industry may well lean with tender care over the cradles of these new born babies like so many good fairies. Their action, however benevolent, will not achieve anything unless, within the same country, there are people willing and able to understand it, improve on it and carry it through.

No one will deny that the training of men is the key to development. But how to achieve this end? That is the question.

One cannot hope to find a ready-made answer by rifling the arsenal of training methods that highly industrialized countries have patiently developed over the centuries. It is wiser to examine the particulars of the problem posed by the new countries in all their originality.

It is well known how discouraging their characteristics are, not only where they concern the number and technological level of those who need to be trained, but also when one considers the diversity of the functions they must learn to perform and the dearth of qualified teachers. Let us examine these three points rapidly:

1. The people who need training almost always belong to a totally different civilization from that of our technical world. Their culture, although far from being unimportant, is the complete opposite of that of the European. Their basic knowledge, often inadequate for the purposes for which they are destined, is rarely homogeneous. Totally unaware of the business ethic, they often even fail to comprehend the rules of the game: why arrive punctually? why try for promotion? why strive after a high standard of work? why obey several superiors? why treat a machine with care?
2. The functions these people must learn and the jobs they must be provided with vary enormously: government employment, mineral extraction, public services, transport, agriculture, etc. Everywhere responsibilities require responsible people. Everywhere there is a dearth of people who have been trained for their jobs and a frightening plethora of untrained manpower.
3. But teaching needs teachers, and this third point is not the least disturbing. There are only a minute number of teachers in the European sense, with doctorates, degrees, or diplomas, and they have generally been absorbed by the existing needs of the state at a high level.

The Search for a Solution

Consequently, if despite these difficulties, one is going to go on with the search for a realistic solution, one must start by analyzing the terms of the problem one is facing. We must consider the problems of teachers, of jobs to be provided, and of people to be trained.

Which Teachers?

In the educational tradition we know in Europe, the teacher is much more than simply the purveyor of information. He prepares a lesson, delivers it, and tests its effects; the preparation seems, on the evidence, to be his most worthy task, the one which requires the most imagination and knowledge.

We will doubtless be forgiven for suggesting that this teacher is slightly similar to the industrial craftsman of the era before mass-production; the craftsman also prepared, fashioned, and tested. His output, usually of very high quality, was tiny and reserved for a rich élite. Today industry has replaced this craftsman by a series of specialists: some to prepare, others to give orders, more to carry them out, and others to test. Those who obey the orders and carry out the operations required are usually the least qualified of these specialists, particularly when the preparation work has been most thorough.

Why should education not react to a problem of mass production in the same way as industry has?

Can we not imagine the men who prepare teaching programs being as thorough as is conceivably possible about producing guide manuals for teachers, preparing teaching aids, coordinating theoretical and practical instruction, scheduling lessons and teaching media and training not school masters, but basic instructors who will pass on these fully prepared lessons to small groups of not more than twelve trainees?

Where then will these instructors come from? Generally from the same background as the trainees, or a similar professional one. They will either simply be chosen from among those who already have the practical knowledge the others will have to acquire, or they will first have to be introduced to the subjects they will have to pass on; in any event, they will have to be taught to perform their minimal teaching duties, for teaching requires less pure knowledge than knowledge of what and how to teach. This gives the first principle of an original solution:

> *Preliminary preparation of the training program, so simplifying the task of teaching, and the use of instructors of a slightly superior level to the trainees, appropriately initiated in active teaching methods.*

Which Jobs to Be Filled?

Let us now turn to the problem of the jobs to be provided. One should notice, not necessarily in criticism, that in an industrialized country training very rarely prepares one directly to fill any specific and narrowly limited job. The successful trainee, whatever the level of his knowledge as confirmed by a more or less imposing document (professional aptitude certificate, engineering degree, doctorate, etc.) feels rather bewildered when given a specific task and is often horrified to find his knowledge superfluous and his ignorance alarming; happily this ignorance is often compensated by an ability to understand developed by many years of attending academic institutions.

Can it be the same for the generations of adults in developing countries whose very difficulty is the lack of basic knowledge? We hardly think so. Everything argues against it for, if one wants to bridge the gap, the rift, the abyss separating the person from the job for which he is destined, one must pass on to him that knowledge and those attitudes necessary for the job, and those alone.

Was this not how some of the most industrialized countries faced their problems of mass education during World War II, when creating rapidly the army they lacked?

The U.S.A. for example multiplied the number of their naval officers by thirty in a few months, recruiting inland men who had never seen the sea to perform specialized officer's jobs on Atlantic escort ships six months later. These men did not claim to have achieved the multi-competence of expupils of the Naval Academy, but they knew how to aim guns of a certain caliber that armed a certain type of escort ship. And they won the naval war.

So one can state a second principle:

Training for carrying out a specific job, encompassing all that is necessary for doing this job and no more.

Who Is to Be Trained?

Let us now open the third shutter of the triptych, the people to be trained.

The long educational tradition in the essentially very integrated human environment of the industrialized countries often leads one to forget the basic principle that one must know the person before one can start training him.

What does this involve? Knowing how much he knows, what interests him and attracts him, how much he does not understand, what he scorns, what he hopes for. One must understand his faith and customs, the fairytales and proverbs of his childhood, his customary values, his technical awareness, the way in which he sees the world.

Let us imagine we have acquired this knowledge of the various candidates for training. We can now split them into homogeneous groups of a dozen pupils. It is then easy to discover how much information any person is going to need to be taught in order to fill a certain job, the gaps not only in his technical knowledge, but also in his practical knowledge and his attitude to the technical world. And the problem thus becomes one of just passing on to him all the knowledge he needs to fill these gaps.

In these new terms, does not the problem become much simpler? For to get to know the trainee we have already studied his cultural bases and we will know his folklore and the images with which he communicates.

And so finally we discover, among the discouraging characteristics of the situation we have already summed up, a promising note of harmony: the instructor who has been picked out from among the trainees speaks their own language in the fullest sense. He will be able to communicate with them better than anyone else and will use the most active teaching methods quite naturally; moreover he will be able to demonstrate at the preparation stage whether his fellows will be able to grasp an analogy or a particular teaching aid.

The strategy of communication, based on using existing knowledge for acquiring new knowledge and on the participation of a "fifth column" of instructors, is not essentially different from strategy pure and simple. This point can be stressed by a quotation from B. H. Liddel Hart's *History of World Strategy*: "When one considers the stages through which new ideas have passed before being accepted, one sees that the actual acceptance process was facilitated when they were presented not as something radically new, but as a revival, in modern terms, of an age-long principle or practice which had been forgotten."

Let us return to our subject to state a third principle:

> *Preliminary study of those to be trained, enabling the formation of homogeneous groups. Preparing the teaching program for each group, with particular reference to the culture of the people involved.*

We have now outlined the characteristics, if not of a general theory, at least of a method of approaching the problem of training.

Taking these characteristics in the right order, we can now establish a pattern for training certain people from a certain background for a certain job, as set out in Table A3-1.

It should be noted that this kind of operation, if it is to be successful, calls for the collaboration of engineers, sociologists, psychologists, doctors, organizers, educators, etc.

A private internationally-oriented company could assemble such a group of specialists and lend assistance to governments and enterprises that wish to set up training centers using the methods previously described. This type of company, specializing in the training, may thus be seen as complementary to companies, which do technical studies and are concerned with procedures and machines, and to management consultant firms, which are concerned with methods. This provides the background for the following account of our experience in developing countries.

Table A3-1

Preparation	
A—Job studies to be carried out	Necessary knowledge and attitudes Selection criteria for future job holders
The men study	Present knowledge and attitudes
The environment study	The men's environment, the work environment
List of the knowledge and attitudes to communicate to the trainees	
B—Selection and training of the future instructors	
C—Training preparation	Division of the knowledge into lessons closely linking theoretical lessons, application and practice, preparation of audio-visual aids based on the men's environment
D—Construction or fitting up classrooms and workshops	Usually makeshift classrooms on the work site Supplementary training equipment to be prepared: tape recorders, slide projectors, occasionally teaching machines
E—Trainee selection	
Execution	
First stage launch	Permits a check on training effectiveness and modification to programmes
Control Checks	
	Permanent written checks of trainees End-of-course examination Subsequent on-the-job testing of employees by supervisors

Examples of Early Experiences

Table A3-2 defines the nature of some training projects that have been carried through successfully in Africa using the method proposed earlier. One should note the variety of techniques and hierarchical levels involved.

Each of these projects no doubt deserves to be developed here in detail, following it through every different stage and observing the results. But without treating each operation in such detail, it may be interesting to mention certain points about some of them.

Case of the Rural Tax Assessors

In the matter of the job-study and the gaps in knowledge this can reveal, the case of the rural tax assessors is particularly instructive. These men clearly have to be trained in agriculture, surveying, and tax law, but we need not dwell on this.

The study of the job enabled one to establish where the gaps in knowledge were, but it also brought to light another, possible more serious, totally unexpected gap. Whereas a few years ago the assessor made his rounds on horseback, he now has a car, usually of a Land-Rover type, which he drives himself. But the roads are often bad, the car is battered and breakdowns frequent. And his knowledge of the finest subtleties of tax law will get him nowhere if he is stranded dozens of miles from the nearest house. It was therefore clearly necessary to include some lessons and some practical training in mechanics and breakdown procedure in the training program.

Case of Total Illiterates

In the matter of studying the men, the case of the illiterates is undoubtedly the most striking. These men were totally illiterate, without even the crudest idea of any language with a developed technical vocabulary, and equally ignorant of writing. They expressed themselves in their mother tongue, various Berber dialects of more or less pure Arabic.

The study showed, in addition to the gaps mentioned above, the complete separation between the illiterate and the technical world. The illiterate lives in a concrete world made up of "things" which he often explains in supernatural terms and not by any rational process. The most simple drawing, or in certain cases a photograph, means nothing to him: if he is occasionally able to draw, it is without being aware of the ideas of vertical and horizontal, of the right angle or of the circle; his spontaneous sketching can have no professional, technical, or educational application; the most elementary technology defeats him: bending a wire to form a simple geometric shape, discovering the direction in which a screw turns and turning it, using a hammer, putting a two-pin plug into a socket, etc.

Table A3-2

Country	Firm's Activity	Job Positions
Tunisia	Iron-ore mining	Lorry & shovel maintenance workers
Algeria	Saharan transport	Heavy-vehicle drivers
	Saharan transport	Assistant drivers
	Saharan transport	Assistant mechanics
	Soil-testing laboratory	Mechanical soil-testers
	Electricity supply	Saharan power station operators
	Electricity supply	Network workers
	Oil exploration and production	Production workers
	Oil exploration and production	Specialized maintenance workers primary education (tech. branch)
Morocco	Cobalt mining	Timbermen and drillers
	Manganese mining	½ maintenance workers
	Manganese mining	Supervisors, underground
	Manganese mining	Timbermen and drillers
	Iron-ore mining	Supervisors, underground
	Iron-ore mining	Timbermen and drillers
	Lead mining	Drillers
	Lead mining	Drillers-blasters
	Phosphate mining	Drag-line drivers
	Phosphate mining	Shipping supervisors
	Phosphate mining	Office clerks
	Phosphate mining	Trainees in a specialized training center
	Mine-transport	Motor mechanics
	Government administration	Rural tax assessors
	Government administration	Urban tax assessors
	Government administration	Bookkeepers
Mauritania	Mining complex involving surface mining, handling equipment, rail transport, stocking and shipping, auxiliary services: electric power stations, workshops, etc.	100 different job positions: driving and maintenance of lorries, locomotives, bucket wheel reclaimers, shovels, diesel-electric power stations, etc.

Trainees' Starting Level	Type of Instructor	Length of Course (months)
Literate workers	Draftsman & European crew leaders	2½
Heavy-vehicle driving license semi-schooled	European drivers	4
Literate workers	European drivers	4
Literate workers	European mechanics	5
Primary education	2 crew leaders: 1 European 1 Algerian	1½
Literate workers	2 crew leaders: 1 European 1 Algerian	1½
Literate workers	2 crew leaders: 1 European 1 Algerian	3
Algerians primary education	European crew leader (in France)	3 + 1 general knowledge
Illiterate workers	2 Europeans 2 Algerians, primary education with professional qualification	6
Barely literate workers	European leader Moroccan supervisor	2
Literate workers	European crew leader	3
Good workers or literate foremen	European overman	3
Illiterate workers	Moroccan supervisors	2
Good workers or literate foremen	Spanish overman	3
Illiterate workers	Moroccan supervisors	2
Laborers or ½ illiterate workers	Moroccan supervisors	1
Laborers or ½ illiterate workers	Moroccan supervisors	1
Literate and illiterate bulldozer drivers	2 European dragline drivers	4
Inexperienced literate supervisors		
Inexperienced clerks, schooling		
Illiterate workers	Crew leaders and Moroccan workers	8
½ literate maintenance men and inexperienced motor mechanics	European crew leader	4 to 6
Assessors, primary to secondary schooling	Moroccan inspectors or assessors	1
Assessors, primary to secondary schooling	Moroccan inspectors or assessors	1
Primary to secondary schooling	Moroccan government bookkeepers	1
Illiterates or semi-illiterates, no industrial experience	European workers with appropriate specialist knowledge, teacher-trained, assisted by educated Mauritanians, without industrial experience	3 to 9 according to particular function

An illiterate's educational deficiency is really a difference of civilization or rather, to use Taylor's terminology, a fundamental cultural difference. Often this difference is accentuated by a misconception of the very idea of progress. He cannot assimilate the constraints imposed by technical progress on his pattern of behavior without requiring further justification.

One is only going to manage to adapt such a person to technical civilization by cushioning him from every shock and taking him slowly from concrete situations and objects connected with the work. This includes accepting his culture as a starting point and using this culture to help him to acquire the first notions of technology, of language, and of writing, and maintaining a balanced harmony as he increases his knowledge of the various domains.

Let us take a simple example: if the trainee learns to use a hammer, he will immediately learn to write the word 'hammer' and will attach a symbolic value to this word as a new center of interest, while an introduction to drawing will enable him to recognize rapidly the picture of the hammer and to draw it for himself. Around the center of interest represented by the hammer and the effects of a blow with the hammer, he will be taught ideas of safety, of rules for using it, etc.

It is suprising to observe how favorable the results are and how much the trainee's interest remains alert.

This phenomenon can be conveyed by an image: the knowledge that the illiterate must acquire can be represented as so many sides of a pyramid; the base of the pyramid stands for the minimal original level of his knowledge, while the aim to be achieved is at the summit of the pyramid. It is easier to reach the summit by a path that ascends regularly successive sides than by taking a direct route up one side (see Figure A3-1).

In certain cases, however, the need is so pressing that this harmonious approach to the acquisition of various kinds of knowledge and attitudes must be rejected in favor of the transmission of bare technical skills to the illiterate. By technical skills is meant all knowledge that is directly related to the execution of a job by the individual; the job is simple and does not require any parallel knowledge of reading or of writing (farmer, miner, dock hand, etc.).

In this case, of course, only one side of the pyramid is ascended; but, contrary to what one might think, this is the least steep side. Actually, the success of any training program depends essentially upon the active participation of the trainee, which in turn depends on his degree of motivation. What attraction could the difficult task of learning to read and write a language possibly hold for a man accustomed to a "natural" civilization? On the contrary, technical skill most often appears to be the doorway to employment, salary, and housing, and an illiterate individual, especially one who has lived for several years in the vicinity of an industrial center, is attracted by the magic of tools and machines.

The motivation is there; the "channel of communication" remains to be found. Obviously, the written word is not the answer, whereas the use of an

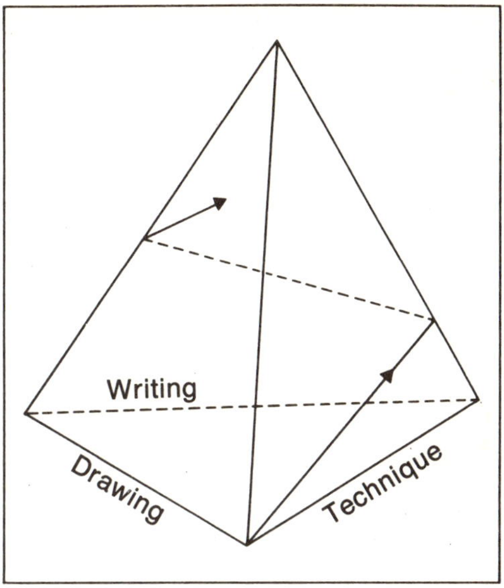

Figure A3-1. Realistic approach to the technical training of illiterates.

instructor whose background and dialect are those of the trainees seems to be imperative.

The solution lies essentially in the use of diagrams and scale models as the basic educational tools, while verbal explanations become mere aids. The first step is now learning how to read the diagrams in a few sessions. The sequence of instruction then follows a strict line of progression as is illustrated by the following example.

Case of Timbermen

Men training to become timbermen are to be taught how to install the framework of a curve mineshaft. The key to this type of timbering is to put in the framework along the line of the curve. The instructional phases are as follows:

Phase 1—Using a scale model, the trainee builds a framework into a straight mineshaft. He has already timbered straight mineshafts at the bottom. With the help of the instructor, he masters this exercise without difficulty and becomes aware of the relation between the scale model and the real thing.

Phase 2—The trainee compares the timbered scale model with the corresponding diagram. Since the scale model is cut away, he can see the whole top part of the timbering. He has no trouble perceiving the simple relation between this view and the corresponding diagram.

Phase 3—The trainee compares the diagram of the straight timbered mineshaft and the diagram of a curved untimbered mineshaft. He is easily able to notice the simple relation between the two diagrams and can then see what makes them different.

The trainee is shown diagrams that represent good and bad examples of timbering in curved mineshafts. As the trainee is now able to "read" the diagrams, it is possible to take a "rational" approach, to show him the correct way of timbering, the main points to keep in mind and the pitfalls to be avoided.

Phase 4—Under the direction of the instructor, the trainee timbers a scale model of a curved mineshaft. Just as he went from the scale model to the diagram of the straight mineshaft, he now goes from the diagram to the actual timbering of a scale model of a curved mineshaft.

Phase 5—The trainee actually installs the timbering in a curved mineshaft. Having learned the main points defined by the diagram for the execution of the scale model, he is now able to timber correctly at the bottom of the mine.

This strictly technical training, of practical value to individual and company, often has more positive results in terms of individual adaptation, than one might think. It actually leads the trainee to become part of a team and provides a sometimes harsh but effective introduction to the various rules of the industrial game. He is encouraged to attend evening literacy classes voluntarily (in order to climb up the other two sides of the pyramid previously described).

A Promotion and Training Plan

The operations listed in Table A3-2 have all demonstrated the efficiency of the proposed method, under which an average of 90% of the trainees have passed the final examinations. However it must be stressed that this eminently Cartesian method has to be applied with the greatest of care: often, one must sweep away both unfounded prejudices and ill-adapted techniques because one is having to embrace "such vast numbers and so widesweeping revisions" that one has to "assure that nothing is omitted" (Descartes, *On Method*).

Criticism can of course be raised. The one that has come most frequently to our attention is: "That's all very well, but aren't you going against man's natural development? By adapting him to a precise job, aren't you adapting him too precisely? And, once he has learned his job, won't he be tied to it for the rest of his life?" We have also been asked: "What about the man's culture? Did you consider that?"

Experience has proved that the first criticism is not really relevant. The decisive step has been taken once the individuals are able to do their jobs effectively and to take up their full responsibility. Then, according to the innate qualities they reveal, and subject to completing further training courses, the better individuals can perfectly easily move up the ladder of the hierarchy. We know of sweepers who, after four months of training, were promoted to become preventive maintenance workers of General Motors' lorries and, after doing this for a year and completing a further four months of training, became mechanics capable of completely dismantling a 200 h.p. supercharged diesel motor, testing all the main parts, deciding on which to replace, ordering the spare parts, and finally assembling the whole motor.

A similar training and promotion sequence was carried out in a large phosphate plant, this time from bulldozer drivers to drag-line operators and finally to site foremen.

This results from the fact that this method of training does not only bestow a strictly specified number of facts, but also techniques of observation, assimilation, and reflection.

As for the second criticism, this is really going to lead us away from our subject, if only in an attempt to define "culture." However, we must mention that if one aims to take an individual from a pastoral civilization into a technical one, to take him across several centuries in a few months, the greatest risk is that of a rupture with his original culture. And so it is not a case of introducing him to the spire of Strasbourg Cathedral or the nature of color in a Van Gogh, but rather of introducing him to the much more difficult notion of the continuity of the world he is coming from with that to which he is going.

To return to the idea of career progression, one point must be made: all training, especially when it is related to promotion, tends to disturb the hierarchical structure concerned, whether a firm, a government department, or a country. This disturbance reveals knowledge gaps to be filled in the lower posts. If this has not been considered in good time, the hierarchy rapidly becomes disorganized, for training can do a lot with time on its side.

So, one is generally led to draw up a recruitment, training, and promotion plan covering a period of several years and requiring a preliminary study of the posts to be filled and the men to be employed.

A Particular Case: Plants under Construction

This plan, bringing with it a dynamic of personnel management, is called for when recruiting and training operating staff for a large industrial plant under construction.

The abundance of raw materials, whether minerals or energy sources, in developing countries is at the root of the construction of important industrial centers, usually using the most advanced techniques and sometimes achieving production capacities that surpass the records of heavily industrialized countries.

All those responsible for the plant's construction will draw up detailed work programs in conjunction with the foremen.

Clearly, creating the crew of the new plant involves just as much care being taken over both detailed preliminary studies of the workers and their job positions and the working out of a recruiting and training plan that will produce, in good time, the necessary staff appropriately trained for their future tasks.

There are many difficulties; as the plant does not yet exist, it is not fully clear what the jobs will be, and operating and maintenance instructions have not yet been firmly settled. However, training can proceed first through the stages of programs common to all workers and then through the various increasingly specialized programs as the construction engineers, the technical specialists, and the future operators establish the exact nature of the plant.

The training is preferably carried out on the same hierarchical pattern, with the foreman, often a European, helped by a local monitor, training the future operators.

This is a broad outline of the process as it was applied in a large mining complex in Mauritania, and foreseen in Morocco for a chemical plant under construction.

Conclusion

Can one make any conclusions after having taken but a few steps along a road as long as that of the training of man? We do not think so. We would therefore prefer to end this article on a note of reflection.

The method we have proposed and tested seems to make the largest allowance for the problems of developing countries. But how does training appear in the heavily industralized countries? There the level of man's education is much higher, its mean value is growing steadily, although this does not stop it from creating serious difficulties for public education.

But over and above this basic education, a new need is appearing, born from the revolution in methods and techniques that one must face during his career, for which no school can do more than "teach him to learn." The rhythm of progress and the ruthless nature of economic criteria are making man the longest-lasting asset in any business. Only a dam has a longer active life (sixty-five years) than man; a boiler only lasts thirty years, a machine tool from five to ten. So he needs to be being continually educated, to change, as Louis Armand, the twentieth century humanist, says, his "framework of knowledge" two or three times in a life-time.

Here we are confronted with a new problem, one which universities and schools have only just begun to deal with, and which industry and government are increasingly trying to solve, at least partially, in their own way. But then they are seeking efficiency: the best adaptation of their employees to their own methods and techniques. They must therefore have experts that will be hard to find outside of their own hierarchies. But then at last they must acquire a deeper knowledge of their workers, whose capacities for learning and improvement constitute a basic asset of their heritage.

Should the three principles we suggested ("The Search for a Solution") be remodelled to solve this new problem?

We think the first and third will stand up well. As for the second, perhaps it is sometimes too restrictive, yet if applied intelligently, it can bring about a breakdown of barriers, an essential step towards culture.

For example, it is easy to observe how difficulties between departments may arise from a misunderstanding on the part of each of the other's usefulness. The exercise of responsibility requires an understanding of the microcosm that is a company; this understanding, in turn, naturally gives access to more ambitious economic topics such as a country's stability.

Thus, it appears that whatever its objectives, any educational endeavor in any country must be increasingly aware of the original pattern of the mind that is to be trained, an awareness that is the basis of the "Copernican revolution in education." This special attention was at first directed to children; now it is the adult's turn to demand from the educator's art what Pascal proposed for the art of persuasion: "Whatever it is of which you wish to persuade someone, you must be familiar with the person himself, and know his heart and mind, what his convictions are, and what he likes."

Appendix IV

Handling A Vast Technology Transfer Operation: Miferma

Original Title: "From Traditional Teaching to Continuous Training" (extract from the Miferma case: Mines de Fer, Mauritania).

Authors: Albert Plent, former deputy general manager of Eurequip, and Silvère Seurat

Reference: *Etudes and Realisations,* June 1966

Summary: *Establishing a huge company in the middle of a desert poses huge technical and sociological problems; the Mauritanian people were not prepared for such industrial expansion. Personnel must be found, teaching organized, locals trained, jobs and organization adapted to local conditions in such a way that the supervisors can take over without disturbing plant running. This is an example of a technology transfer success.*

Engineers will be impressed by the achievements of Miferma (Société des Mines de Fer de Mauritanie) and by the many records that have been broken: a huge mountain of iron ore to be tapped, over 400 miles of railway constructed across a hostile desert, used by 10,000 metric ton trains made up of three engines and 2,000 yards of trucks, with a capacity of 4,000 tons an hour both to the yards and back and while unloading trucks and loading ships, two boom towns, two 10,000 kw power stations, all in very unfavorable weather conditions: sand and wind everywhere, sea-level humidity at Port-Etienne and extremes of temperature at Zouérate.

Despite these mighty achievements, Miferma still had another problem of an equal scale to face. Nowadays, thanks to sociology, it can be solved. They wanted to employ as largely a Mauritanian labor force as they could. So in a few months many people had to be shown the way from patriarchal to technical civilization, a journey that took us several centuries.

This problem was the subject of a preliminary study that led to a plan for training. A consultant engineer and a psychologist carried out this study a year before the first train-load of ore was due to run. The tasks required by the extracting and transporting process were analyzed and expressed as definite jobs to be filled; the nature and origins of the labor force were also studied. The conclusion, put simply, was this: over 500 virtually illiterate workers, totally unaware of any basic techniques, must be trained for 50 different jobs.

Previous experience, notably in the Sahara and in Morocco, has shown that training is most effective in small work groups, each with a maximum of ten trainees and directed by an instructor using active teaching methods.

Where were the fifty instructors to be found? There seemed to be only one possible answer: pick them from the future foremen. This is certainly an inconvenient and risky solution, for it means that the future instructors have no teaching experience and usually know more about the tricks of their trade than about its logic. But on the other hand, this answer is extremely satisfactory as it enables the establishment of channels of communication within the future working structure.

And the instructor learns to lead a team of Mauritanian workers at the same time as he realizes the scale of the revolution he is imposing on his trainees, while helping them to sustain it quite naturally. But let us return to the difficulties; they were quite easily overcome.

The instructors were given one month of teaching training: they were taught the pattern of production they would have to operate as well as the characteristics of the people they were going to teach. They were also integrated into an important part of the preparation of the training program during which the syllabus for the various lessons was fixed in very great detail. This meant that the improvised instructors would teach all that was to be known, without mistakes or omissions, and also use a language and arguments which their pupils could grasp.

One of the most important characteristics of the "school" that Miferma organized for one academic year, as well as the abundance of pupils, instructors, special textbooks (about 35 textbooks with an average of 250 pages) and teaching aids (charts, models, etc....) was the almost total absence of traditional school buildings. Everything took place in huts on the construction site, in the atmosphere of the site: it was an operation like any other, but the aim was not to create a building but a crew.

On these terms the challenge was accepted and the men were ready at the same time as the plant, that is to say, and this is rather unusual, several months ahead of the official program.

Miferma and Eurequip then set out on a second stage. Some of the Mauritanian workers who had had this training had shown in their work that they were clearly promotion material. It therefore seemed desirable to prepare a pattern for promotion (see Figure A4-1). But one had to take into account the complicating factor that the firm's structure had to be adapted to the characteristics of the two labor forces it contained: European and Mauritanian.

In fact, the jobs held by the Europeans, which had indeed been created for them, were made up of groups of tasks of widely varying complexity, of which some were beyond the range of a Mauritanian. It seemed, therefore, as if he were facing a promotion barrier. The key to promoting a Mauritanian was to modify the job structure so that a series of stepping stones obviated the need for this barrier.

Two examples can be quoted to illustrate the procedure we used.

In the mine extraction equipment, the four shovels, Bucyrus B electric shovels, were operated by European technicians. Each shovel's seven-cubic-meter capacity means it can extract a thousand metric tons an hour and represents an investment of six million French francs. Mauritanians were gradually promoted after learning stages on increasingly difficult equipment. They start by learning to drive a tanker, then go on to a Dumper 27 T, a Dumper 65 T, a bulldozer, a mechanical Marion 93 M shovel, the Bucyrus mechanical 88 B shovel and finally the electrical 150 and 190 B shovels. At this stage, about five years after his training started, a Mauritanian can become a junior instructor.

In the maintenance department, the Dumpers section develops in a similar way through a restructuring jobs. Developing job organigrams are drawn up

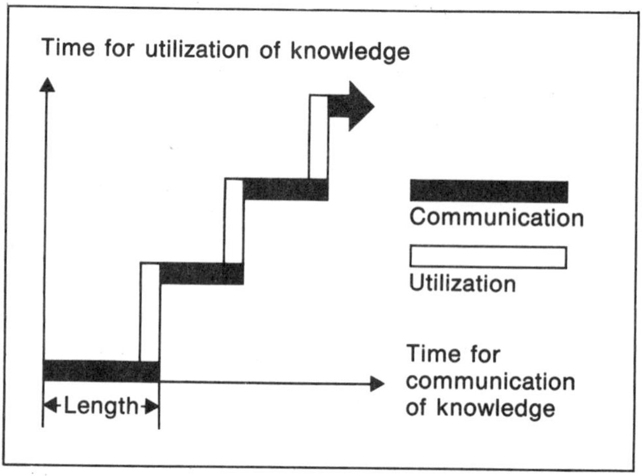

Figure A4-1. Promotion cnannels for Mauritanians, alternating communication and practical application of knowledge (Miferma).

between the initial jobs structure and the target structure, so that new tasks can be given to the trainees each time they reach a certain level of competence.

So it became possible to tabulate a graded time-schedule in terms of experience and training for the promotion of a Mauritanian and to plot a five-year training plan.

This is certainly possible, but relatively complicated, since the rules of industry require one to stipulate a cost-effective optimum. By processing this data with a computer, Eurequip was able to determine the optimum time schedule for this, as well as others for various hypotheses of expansion of business and the evolution of a stable labor force.

Psycho-Sociological Lessons from the Miferma Operation

During a conference organized by Eurequip in 1967, we were able to imagine the precise conditions in which this successful and "enormous mutation" took place, when one of our engineers evoked in the following terms the essential comprehension of Mauritanian habits and customs.

When a technician from an industrialized country arrives in a job he unconsciously brings with him an ethical tradition, resulting from wide-ranging experience, which permits him to apply to his job the efficiency expected from him. It is natural for him to be punctual, to stay on until his work is finished, or, on his own initiative, to look for other work to do during slack periods.

But the Mauritanian does not possess this second custom—the efficiency and behavior that would aid his industrial integration. Given the liberty he takes with his own time during his daily life, the Mauritanian can be easily surprised by the fact that the bus driver does not wait for him. Directly attached to his immediate boss he does not easily make the distinction between the person and the job, nor does he immediately obey an order from higher management that does not pass via his usual superior. He is completely insensitive to certain signs that indicate serious problems with machinery: noise, overheating, and so on. Slack periods are not an opportunity for doing secondary work, but rather for leisure, opportunities for past time games while waiting for more work to be distributed. The Mauritanian does not easily accept transfer from one job to another and one observes that he puts machinery into an order of importance according to personal criteria not shared by the company.

This relationship between the Mauritanian and his job is not uniquely Mauritanian and can be found in the West too. What is uniquely Mauritanian is that possible solutions are not helped by the local basic culture and each time one must study the overall situation, weighing possible solutions against industrial and cultural criteria. In any case, the solution is never found in a witty remark, by a good talking to, in a shrug of the shoulders, or an appeal to good-will. The solution is always slower and more complicated.

Outside these differences in their attitude to work, we have encountered deeper attitudes that involve their view of existence in general. The Mauritanian belongs to a basically religious culture. He is imbued with certain fixed ideas concerning his destiny; he follows precise rules that regulate his life from birth to death, whereas the European comes from a civilization firmly anchored either in methodical doubt or liberal scepticism that results from the wearing away of political, religious, or social explanations of his destiny or reason for living.

As one would imagine there is no immediate point of contact between these two peoples. The understanding established between two people from the same background, the understanding that goes beyond communication by language does not exist (certainly at the beginning) between people from fundamentally different backgrounds. Only time can establish a spontaneous, nonverbal communication. We have also encountered difficulties in getting across information, instructions, and orders.

To arrive at a minimum of communication language classes must set up, and a means of explanation must be found that is understood in the same way by both European instructor and Mauritanian trainee. For example, to make some Mauritanians understand the mechanical effects of friction the reference to skin on a camel's foot being rubbed by the hobble has been found illuminating. To make them understand the principle of a compass tracing on sheet metal, reference to the tracing of the surface of a tent has brought rapid results. These simple examples indicate the need for educators seeking communication, to find common elementary examples with a mutually understood interpretation. That is valid not only for psycho-sensorial understanding, but also for more elaborate phenomena involving wider ranging operations.

Appendix V

Is the Training of an Industrial Workforce Profitable?

Author: Silvère Seurat

Reference: Conference at the International Bank for Reconstruction and Development, Washington 1968.

Summary: *There is a definite link between investment and the need for training; but the solutions generally adopted to solve the training problem do not have the same value. It is possible to reason objectively and to consider only economic factors. The global cost of training in an undertaking can be evaluated through measurable parameters such as investment, finished products...and thus to calculate write-off. The results obtained demonstrate that training can be profitable.*

My presentation will be linked to two main ideas that will be found in various examples I will put forward. The first main idea is *the connection between the investment of capital and the need for training*. The second is *the profitability of training*.

The Link Between Investment and Training Needs

It is obvious that any investment generates jobs that are specific to this investment. According to the nature of the latter, the jobs proposed will be more or less complex; they can be laid out on a scale of growing complexity, from the floor-sweeper to the astronaut.

In a highly industrialized country, the impact of a fringe investment on the employment curve may well go unnoticed as long as the degree of complexity is not too high (see Figure A5-1). The number of men available on the market is sufficiently flexible to meet this demand.

This is no longer true when a certain degree of job complexity is involved or when the employee has a certain amount of economic responsibility. For instance, operators of an oil refinery are selected and trained two years before production starts; this period is extended to three years in the case of a nuclear power station. In this way also, the seven American astronauts chosen from among the best military test pilots, were recruited and trained two years before the first ballistic flight made by Alan Shepard, and three years before the first space flight of John Glenn.

A "critical point" appears where the training curve cuts across the employment curve; beyond this point specific training is desirable, if not imperative. In a developing country, this critical point is reached at a very low level in the scale of complexity of job positions. Industrial experience is practically nonexistent, the available manpower has received school instruction only, and the mere opening of a mine upsets the precarious balance between training and employment.

We have noted that, in such circumstances, various solutions can be, and are in fact, applied.

Investment is entrusted to local people who are ill-prepared for such a task. Equipment is frequently damaged and becomes unusable within a very short time (this was the case in Algeria in the agricultural field).

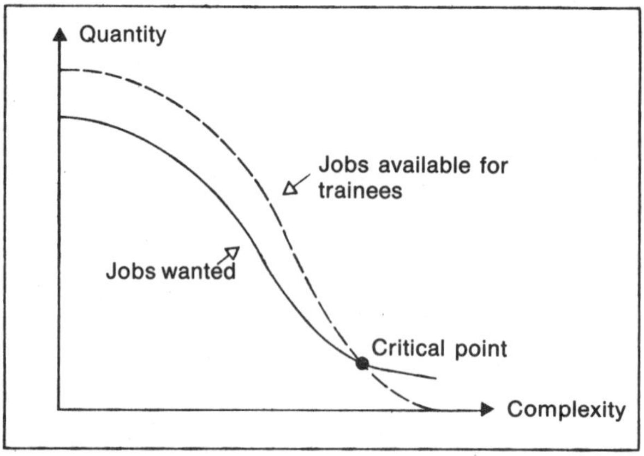

Figure A5-1. This graph shows the typical relationship between job quantity, complexity, and availability. When the "critical point" is reached, specific training becomes necessary.

Foreign technical assistants are installed; they are numerous and skilled. This technical staff runs the plant, but keeps the local people in the background without caring about their training. The number of these men increases as they have less real responsibilities. For this reason, the financial equilibrium of the concern is made precarious and the future is insecure.

Technical assistance is provided within the scope of a general training schedule, enabling local people to gain industrial experience and repatriate the technical assistants without adversely affecting the operation of the concern.

Profitability of Training

At this point, I would like to limit the scope of my remarks and state clearly that it is not a question of determining the benefits derived from a general education program for the whole of a country.

Our project is more modest and practical; it simply consists of evaluating the benefits of a training scheme for the crew of a new plant. For this we shall start with the local working population as it exists at the time of investment. Here it is convenient to consider the diagram in Figure A5-2.

We shall presume that our plant is concerned with a certain "product,"either electric power (kwh), transportation (tons/km), road construction (kms), fab-

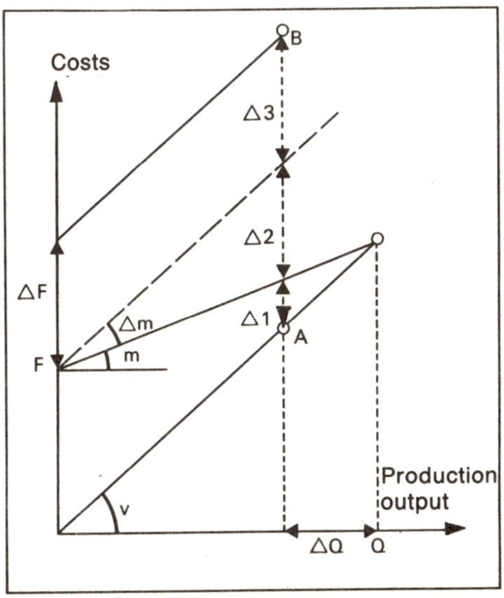

Figure A5-2. Calculation of training costs.

rics (in metres), etc. In good conditions of organization and operation, the yearly net profit accounts are expressed by the following:

Q= Production output
F = Fixed expenses
m= Proportional expenses

Products are delivered to the sales department at a hypothetical price, V, so that there is no profit at the production stage (profit being made at sales level):
Result $= Qv - (F + Qm) = 0$

This is the case in normal conditions, although it is well known that these normal conditions are not reached during the first year. Differences will appear:

$-\ \Delta Q$ (in production)
$+\ \Delta m$ (in efficiency)
$+\ \Delta F$ (in fixed expenses)

The result is a loss, equal to AB, or:

$$AB = \Delta 1 + \Delta 2 + \Delta 3$$

$$= F\frac{\Delta Q}{Q} \text{ (difference in quantity)}$$

$$+\ Q\Delta m \text{ (difference in efficiency)}$$

$$+\ \Delta F \text{ (difference in fixed expenses)}$$

and it is clear that the degree of personnel training, the degree of fitness of men to their functions in the plant will influence these differences:

1. Difference in production: depending on the incidence of breakdowns.
2. Difference in efficiency: depending on the degree of wastage and losses.
3. Difference in fixed expenses: depending on number of technical assistants, number of untrained and ill-employed local people, and local turnover, vocational training being an eventual factor in integration.

Calculating Profit Returns

We will of course calculate the relative profit returns by comparing the two solutions: on the one hand the organized training solution, and, on the other, a reference solution, i.e. one which would generally be used in the absence of organized training.

In considering each solution, the following points should be brought forward:

1. The additional expenses, which are relatively easy to calculate accurately by specialists: expenses covering teaching program preparations, instructors' and trainees' wages, etc....

2. The additional returns, as combined variations in output, efficiency fixed expenses.

It must be admitted that, as opposed to the inevitable expenses that will be incurred, the returns will often depend on uncertain features, particularly those returns pertaining to production output or efficiency. Training thus appears as an insurance premium and it is often easy to demonstrate how, by reducing certain variations, this premium can be amortised within a given period, one or two years for instance.

It would not, anyway, be wise to work out profit returns on the basis of the investment's lifetime or on the individual's lifetime. On the contrary, the calculation will be limited to the first few years of operation, i.e. from one to five years, as the case may be. It is then easy to determine the point at which the investment is recouped, and beyond which returns accrue. The application of the discount rate has no great bearing on the results, considering the short period.

According to the circumstances, the reference solution may imply the following:

1. Resorting to a numerous staff of technical assistants. The calculation is then limited to the fixed expenses variation.
2. Resorting to a smaller staff of technical assistants, which implies the hasty hiring of an ill-adapted local personnel. The calculation is then based on the three variations.
3. In a developed country, resorting to the labor market resources while giving brief complementary training. Variations in production output and efficiency are then considered.

What results can be recorded? In France, by avoiding one week's interruption in the first year of an oil refinery's activities, one can offset the cost of training.

In an electric power station we get the same result with an additional 5% of nonactivity (two weeks) and 5% decrease in consumption (100 kcal./kwh). In a developing country, the investment is amortised within a period of two to four years of operation: two years for a large mining complex or a highly automated factory; four years for a large mechanical engineering plant.

We can see from these results that the cost of human error increases with the present trend towards huge industrial plants, although manpower does not change much. This is shown by the natural tendency towards an increase in the benefits derived from training.

In conclusion, we can state that investment in training, within the scope of a large project, is now one of the fringe investments assured of the greatest returns and that this tendency should grow with the increased size of industrial plants.

Appendix VI

An Example of How to Use a Self-Training Organization in the Creation of an Advanced Technology Industry

Original Title:	"Training Personnel for the Arzew Liquefaction Plant"
Author:	Albert Plent, joint managing director Eurequip
Reference:	*Etudes et Realisations*, May-July 1969

Summary: *How can the maximum number of jobs be given to local workers in the case of a very sophisticated plant starting up in a developing country? The Camel case, Algeria, illustrates and crystallizes the method used. It is based on job reorganization so as to create career channels. Starting with simple and relatively easy jobs the worker moves upwards by routes adapted to his aptitudes and undertaking's needs. The aim of permanent, organized worker advancement extends up to high-level jobs.*

Building a gas liquefaction plant is one thing. Putting together a workforce to run it in the best conditions is another; special training methods must be applied.

Liquefying Natural Gas from the Sahara

The gas is liquefied, reducing its volume 600 times, then shipped at a temperature of minus 161 degrees centigrade at atmospheric pressure in specially constructed tankers. Two British methane tankers *Methane Princess* and *Methane Progress* ship 1000 million cubic meters a year to a regasification plant at Canvey Island at the mouth of the Thames, for the British market. A French methane tanker *Le Jules Verne* supplies the French market with 500 million meters shipped to Le Havre.

Three liquefaction plants, covering 200 acres, can produce 2700 cubic meters of liquid natural gas (LNG) daily; they include peripheral units (for dirt removal, cooling fluids production, stocking) and loading installation capable of loading a 2500-cubic-meter tanker in 15 hours.

A Pilot Plant Handling Dangerous Products

The plant is unique, a million-dollar prototype for methane liquefaction, implementing an original process and new transportation means. Ethylene, butane, and methane flow through an impressive network of pipes, compressors, and columns. Permanent danger makes security a prime concern. One is reminded of this by the ever burning flare illuminating the bay of Arzew.

Staffing a new plant with a crew capable of starting and operating the equipment, and working out and implementing safe and economic running procedures is never an easy task to cope with. In this particular case, it was a real challenge because of

1. Low temperature and high pressures
2. Novelty of the process
3. Lack of manpower with suitable industrial experience with refining plants and automated plants

In order to meet the requirements of the Algerian government and to reduce costs, as many posts as possible had to be given to local personnel. Further investigation revealed that the cultural standard of the people of Oran was high enough to provide the plant with a sufficient number of applicants with a satisfactory educational background, but lacking technical and professional experience in similar fields, which would have enabled them to understand the processes used and the equipment operated.

Advertisements in local newspapers produced 100 to 1000 applicants out of which seven were selected for a training course. The search for production operators turned out to be more rewarding than the search for maintenance mechanics. In all cases, undergoing training courses in advanced techniques was a strong motivation. It is this training that we are going to deal with.

The Training

It starts soon after European personnel have launched the plant, when it is not quite ready to operate. The typical operator to be trained is a 22-year-old Algerian, he has attended a technical school, has a certificate of professional aptitude, and has had various unskilled jobs for two or three years. In turbine control stations (on the central unit or liquefaction units) the operator is responsible for $4 million worth of equipment. In a liquefaction unit's control station any mistake causing a release costs $40,000 for every day the unit is not operating, even if there is no material damage. Even with qualified European staff, releases are frequent.

How can future operators be given a good knowledge of the entire plant, including the processes and procedures to be followed on the job? Moreover, how will they become aware of their responsibility to concentrate exclusively on their job and acquire sound judgment in order to ensure the safety of production?

The answer to these questions lies in the development of training channels. Figure A6-1 illustrates the channels in the production sector.

In fact, the positions of operator can be classified according to the increasing difficulty of the job and the actual responsibility of the person in charge. This responsibility can be assessed by considering the cost of possible stoppages, damage sustained by equipment, or personnel accidents.

So, training will be organized in a stairway pattern, each successive step leading to a more difficult job involving heavier responsibilities. At the end of the first training course the operator is able to cope with the easiest jobs. He is made responsible for it in order to enable the plant to derive benefit from its training investment and to reward the trainee for his efforts.

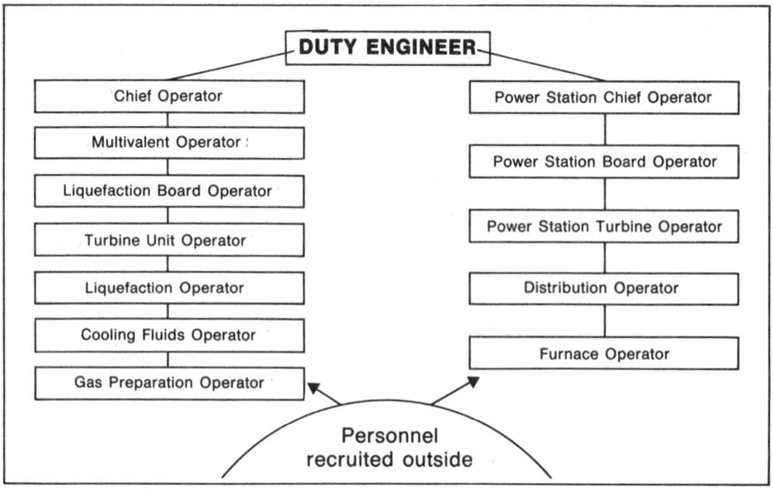

Figure A6-1. This diagram illustrates the development of production training channels.

When he achieves mastery of his job he will be ready to undergo further training and to climb a new step; this solution provides numerous advantages. By way of illustration, let us consider a person who is promoted to the board operator's position:

1. Obviously, following the training channels give him complete knowledge of the section he manages and a keen sense of his responsibilities.
2. Moreover, having a chance to apply rapidly the knowledge he acquires while undergoing further training improves the benefits derived from training.
3. Lastly, he becomes multitalented, which is very helpful for the engineer in charge of production because it makes unit operation more flexible.

The first group of trainees was recruited in May 1966 and had reached board operator's position by 1969 after 24 months on the job and 12 months of training distributed over four training courses. Of course, the training channel is also a promotion channel in order to motivate the men.

The next thing to do is to make an estimation of the qualification level of the job position and of the corresponding salary according to training duration and the responsibility involved.

How the Training Is Carried Out

After designing a general training plan, specific training problems have to be solved:

1. What sort of knowledge should be made available to future operators?
2. Where can instructors be found?
3. Will trainees be taught gas physics and chemistry as well as temperatures and thermodynamics?

With this approach, there is no definite answer. Of course, one would like the operators to understand what they do. Moreover, operators will want to be promoted, which would require them to have a wide theoretical background. On the other hand, many excellent European operators have a very limited theoretical background. The solution to this problem is to determine what knowledge is to be included in the operating instructions, which are the goals of the training action. Those instructions describe accurately what should be the operator's behavior on his job.

Of course, it is ridiculous to turn operators into robots by training them to apply instructions they would not understand. They must be taught what is necessary to understand what is really going on. This can be done by grouping and teaching the theoretical knowledge related to each operational behavior while ignoring what is not derived from on the job practice.

Instructors are chief operators or multitalented European operators in the plant. They have to prepare the courses they will teach. This means that:

1. Operational instructions must be collected or sometimes worked out.
2. Management personnel are responsible for elaborating these instructions for they are responsible for the operation of the plant. Moreover, it is crucial that the language used during the training be the same as the language used on the job.
3. A complete scenario for the training must be derived from the operational instructions according to the methods of institutional technology such as active training, training adapted to trainees, inductive procedures, implementation of instructional media, and simulators.

Expensive simulators are not required to train operators. A simple one, with a panel drawn on cardboard, using very simple electrical connections will suffice. The reason for this is that the operator's procedure is not of the sensorimotor type but of the intellectual type. It is no use training him to perform difficult manual operations requiring rapid responses; he must be taught to make a sensible choice of information and to use it. This is why it is important to show the trainee a board panel showing its connections with outside operating equipment.

This can be done very rapidly. Instructors get a brief training in instructional methods. Thanks to their preparation and teaching work, European instructors entrusted with the responsibility of the training operation for a few months acquire a very good technical knowledge and become real supervisors. There are two basic job positions: gas preparation operator (for the liquefaction channel) and furnace operator (for the utility channel).

Training for these positions is organized in the following way:

1. A one-week preliminary course on the units gives future operators some idea of the equipment they will have to operate and enables them to understand the new words they will hear in the plant.
2. Three months of training with both technical training and practice on simulators.
3. Six weeks of cooperation with the incumbent operator.

In order to make up for possible failure or drop out, six or seven trainees are trained for five jobs. Should a switch from one job to another be possible, further training is grouped in a similar pattern, only shorter. There may be one trainee at a time with one instructor supervising different men being trained for various jobs.

Under normal running conditions this procedure is more flexible. It can be set up thanks to a method of planned self-improvement. Each trainee is given a handbook for every session. During the session the trainee is assigned several tasks that he will have to perform on the unit he will be working for. There may be diagrams to be drawn, equipment to be analyzed, or texts to be studied. Then, the trainee has to answer control questions from the manual and to keep a record of the questions he leaves unanswered in order to ask the instructor for the answer.

Training in Maintenance

Training for maintenance was planned on a similar pattern. Operational instructions are replaced by operational scales and training channels. Training channels are also used to promote technical agents in the plant to instructors' positions.

As an illustration, Algerian instrumentalists have been trained through the following channels:

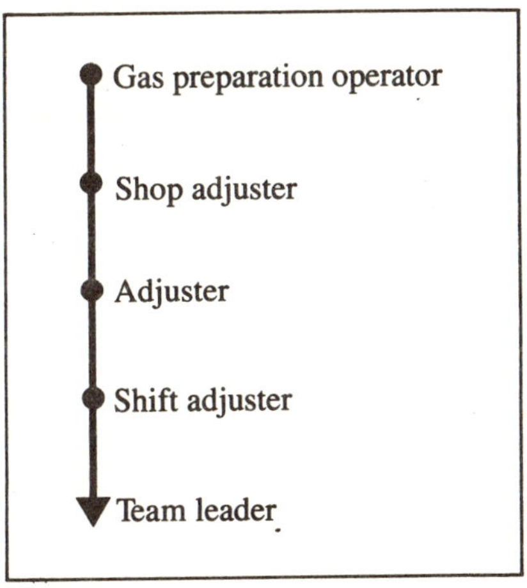

Training from operator's position to adjuster's position takes six months. An Algerian instructor was entrusted with the job, and the results were excellent.

Appendix VII

Application of Transfer Methods to the Strategic Behavior of Companies

Original Title: "Organization, Communication, and Company Plan"

Author: Pierre Davous, deputy general manager of Eurequip

Reference: *Planifier L'Entreprise,* Chapter VI, Editions Marabout s.a. Verviers, Monde Moderne, 1974

Summary: *An educational gap appears when a company starts strategic planning. How is it to be filled? At the level of language and behavior the gap is considerable. The formula which bases action on external courses affecting a few people is ineffective in the majority of cases. We must influence both the men and their environment to give them an incentive to progressively learn the structure; this learning is based essentially on real problems facing the company. This appendix examines some principles of training for change.*

An Essential Cultural Transfer

A company beginning strategic planning must bridge a wide cultural gap involving language as much as behavior—the two fundamental components in any culture.

Language Obstacles

The meaning of words in any language is colored by the emotional interpretations of those who use them. Such a "word drift" involves wide differences from one nation to another. Where words relating to action are concerned (plan, objectives, management) Anglo-Saxons seem to color the words with ideas borrowed from economics and technology. Latins seem to modify the meaning of action words with ideas from politics or sociology.

In a political sense the word "plan" evokes ideas of *dirigisme,* state planning, antiliberalism, and will even be described as a left-wing idea. The word "strategy" is a strongly associated for some people with militarism.

Many mental blocks can thus be thrown up by the use of certain words. On top of those individual blockages we find collective blockages: in a company that has unsuccessfully tried to apply management by objectives we must beware of negative reactions to the word "objectives."

An yet we must use words. In the plan's initial stages—even if it seems a waste of time—it is a good idea to organize an open debate to arrive at a consensus of the vocabulary to be used.

Behavioral Obstacles

These are even more serious than language obstacles and less quickly solved. Without going into an exhaustive analysis of undesirable attitudes and behavior found in companies, we will center on three characteristics that tend to develop among employees in some companies as a result of the protective company cocoon:

1. A tendancy to shy away from risk-taking and to prefer routine activities.
2. A tendancy to shy away from conflict and to remain passive and dependent.
3. A tendancy not to face up to unclear situations and to make superficial, intellectual analyses of situations rather than identify at any cost opportunities for action and to vigorously exploit them.

However, these three tendancies are the three major inhibitors when it comes to adopting positive attitudes to planning and above all, strategic planning.

Planning in fact makes us look at the future in order to decide today's actions. In this leap into the future we will be led in a clear, open, and formalized way to:

1. Make choices despite much uncertainty; and as each choice means other choices must be abandoned, it leads to frustration and is demotivating.
2. Analyze not only strengths, but also weaknesses—and people do not like admitting weaknesses.

3. Take risks, demonstrate aggressiveness, which presages the future conflicts we try to avoid.
4. Face up to the problem of personal capacity (Will I be capable of successfully managing all that?) when we are not certain that God is on our side and success assured. That uncertainty is the most daunting of all. (Part of this analysis is borrowed from a lecture by Ed Bartee, Nashville Graduate School of Management and suggestions from Dr. M. Filiu of Eurequip.)

These attitudes, which seem widespread and valid on both sides of the Atlantic, are complemented in Latin countries by a particular attitude that goes by the name of "system D"—"D" standing for *debrouillardise,* resourcefulness. All formalized preparation for the future is rejected and choices are made on the spur of the moment. The flexibility, speed, and efficiency of such an approach is frequently mentioned, together with its successes and we must acknowledge that such a view is not totally unfounded.

How Do We Bridge that Cultural Gap?

What management information will bridge it? Among the majority of companies the answer to that question is pretty useless. "All you have to do is send top management on a planning course or on a training seminar..."

It's not so simple. Take the example of a group of top managers sent by their company to a planning seminar. In most cases rather than reducing undesirable attitudes it reinforces them, as the following illustrates:

1. We want them to accept conflict; but they find security in the return to the classroom, and the comforting dependence on the tutor leads to passivity.
2. We want them to leave routine behind and accept risk and the anxiety of the head-on-the-block decision. They'll attend the classes and, at best, study case histories that do not commit them. Without underestimating the value of business games, we must recognize that the impact of a game failure is far less impressive than the long-lasting consequences of wrong decisions in real life.
3. We want them to face up to unclear situations creatively, constructively, and positively; but the seminar will immerse them more in learning than in action. The brevity of the seminar demands that much of the information dealt with, even in case histories, is theoretical.

But before anything else we want them to change their behavior. Are they motivated to do so? And is not the seminar seen as a manipulation by management?

Evidently another method must be found—and our criticism suggests it. A learning period must be introduced into the organization. But because what must be learned touches behavior in action, the learning must be based on real problems facing the company. The knowledge and behavior acquired will be driven home by analyzing the consequences of actions following the first learning phase.

Yet, because we cannot learn everything at once, we must proceed by stages with a gradual build up. From that follows the concept of permanent training dealing with real planning problems as opposed to theoretical training.

Principles of Educating for Change

When it is a question of training through action, and thus of inducing new behavior, we believe that the only really efficient approach is one in which the trainees solve real problems. That recommendation applies not only to individuals, but to groups too.

The approach starts with an evaluation of the group's educational level and seeks to raise this level by increasingly complex stages. Each stage starts from real situations understood and mastered by the group; these situations are presented in the form of a model easily understood by everyone. Working with the model one develops a degree of abstract objectivity and this new understanding is immediately applied to real situations.

Methods using deductive and inductive logic must be closely married. It is also necessary to include behavior (involving the group, motivation) and not to stop at intellectual advances.

Training for Strategic Behavior

By applying the recommendations mentioned in the preceding section we can summarize the approach that permits the company to control its strategy.

We must induce a change in behavior. So, in-company groups will work on real strategy problems. And one of the major strategic activities is the creation of the strategic plan. Thus, we can use the preparation of the strategic plan as a real educational problem. The result is that the first version of the plan (and perhaps a large part of the second) has an educational aim. And it is accepted as a condition of the game that the plan may be imperfect at this stage.

Management announces its intention to start a planning operation and to use the plan as a means of advancing the company. It clearly explains the educational character of the first planning stage and the relatively small number of people involved in planning. The explanation is given orally to managers, the works council, and the board.

The working groups are designated; they include higher management, as well as young middle managers whenever possible. During the first planning stage a fairly unsophisticated work schedule is chosen, for example:

1. A brief discussion of the plan's nature and vocabulary
2. A study of the business environment (opportunities/threats)
3. A study of the business potential (strengths and weaknesses)

4. Some thoughts on the targets
5. A study on the interaction of external factors (opportunities/threats) and internal factors (strengths/weaknesses)
6. Identification of the major strategic directions
7. Relationship between strategic directions and targets
8. Feasability test (relationship with available means)

Throughout the first phase (which can be finished fairly quickly—six to ten meetings in one to three months), we are careful to avoid perfectionism and losing ourselves in details;

1. Keep a record of logical links discovered during the various discussion stages, links that cannot all be taken into account during the first planning meeting
2. Identify certain obvious points that can improve the strategy or to improve the running of the company

No matter its imperfections, the result of the first phase will be presented to top management. Oral presentations will be made by the planning managers at several different management levels, arriving finally at the works council. All reactions will be noted and carefully studied.

The second planning stage now begins. Managers involved in the first phase are invited to undertake a second study at a higher level.

It starts with a critical analysis of the first planning stage, the emphasis being put on the relationship studies. At the same time, a first phase will begin in each major division of the company aiming at greater elaboration of the management plan. The number of people involved will, therefore, be much greater than during the first planning phase. At the end of the second stage we arrive at a second version of the company plan and a first version of the management plan.

Here again, the number of management levels to be informed must be carefully defined. In a medium-sized company all managers are usually informed orally on both the company and the management plan.

The rest of the process is clear enough: in each sequence the managers already involved consider increasingly complex problems. Meanwhile, new groups of managers are involved for the first time.

Ideally after three or four years all decision-making managers should be involved and all managers and foremen should be informed of what is happening.

The planning process is enriched at each stage: for example, strategic watchtowers for the companies' business environment are established and contingency plans that describe a future scenario and the action of the company should take should such a scenario become reality are made.

We can represent this stage process diagramatically with Figures A7-1 and A7-2, Figure A7-1 represents the gradual rise in the analytical level in which the planning group works during the various stages; and Figure A7-2 represents the penetration of planning behavior in the company.

At each stage the planning group must carefully study and help to define

1. The optimum analytical level for each planning group according to its educational level.
2. The logical interfaces between the work of each group, without sacrificing flexibility to logic.
3. The information that must be given to the whole organization on the progress of the planning operation.

The group must be aware of the psycho-sociological environment of the planning operation and urge management to call periodic information meetings, which are essential to maintaining a favorable climate.

But we repeat: the group's first responsibility is to prevent the program from degenerating. At the beginning the plan's content is only a secondary consideration.

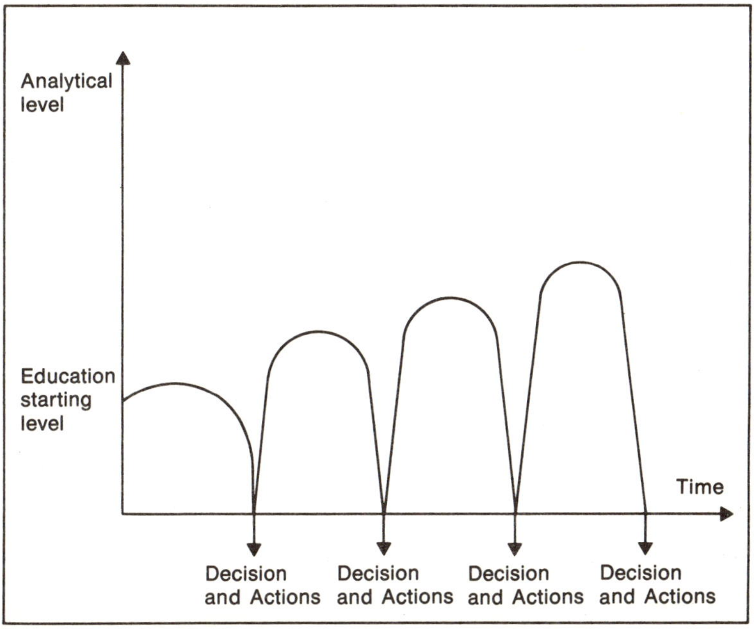

Figure A7-1. This represents a planning group's gradual rise in analytical level.

Figure A7-2. Progressive penetration of strategic behavior in a company.

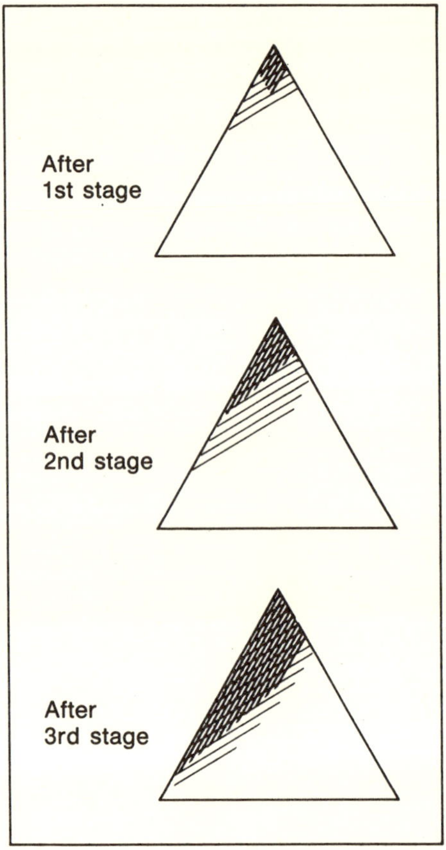

Experience shows that while this watch-keeping task is fairly light during the first planning process, it rapidly becomes more important as the number of groups grows.

From the beginning of the second phase a calendar must be drawn up—with the agreement of the operational managers—showing the work sequence of the various groups and noting that the work conforms with the timing of other important events that mark the life of a company. For example, close attention must be paid to publishing the plan just before annual budgeting starts so as to aid the preliminary budget steps.

Arriving at the end of this exposition, we hope we have encouraged the reader to think about the role played by the communication and planning systems in the social equilibrium of a company. It is of major importance that we teach all members of the company to communicate. It is more than twenty years now that management writers have been insisting on the importance of communication and an open dialogue to maintain a good social climate in the company. It is of major importance that we teach all members of the company to communicate and to ensure that the communication system does not break down. Moreover, the kind of information that is circulated must be tightly defined.

Social developments in recent years clearly indicate the need to open a dialogue in companies on strategic problems and operational planning.

Far-reaching mobilization of company brain power to think about the future of the company is the most solid basis for a management system because this approach improves the coherency of actions and the cohesion of management teams.

Without seeking to impose a monotonous conformity, one can nonetheless hope to see the variety of individual talents and projects harmonized with the overall aims of the company.

Bibliography

"A Conclusive Experiment: Fewer Accidents, Better Service," (Description of an assignment with a large trucking concern). *Bulletin des Transports,* 1325, March 1967.

"A Facet of Sidbec Project: The Transfer of an Iron Metallurgy Tradition," *Etudes et Réalisations,* 107, July-August-September 1966.

Archier, G., Ardisson, E., and Puyo, J., "The Regrading of Workers at the Adour Ironworks," *Le Boucau, Délégation à l'Aménagement du Territoire et à l'Action Régionale,* January, 1967.

______ and Peuchmaurd, M., "Data Transforming," *Hexagone,* 59-60, October 1969.

Bellier, L. and Consigny, V., "Men in Business and in the Environment," *l'Industrie du Pétrole dans le Monde,* January-February 1975.

Brahimi, M., "Technology Transfer," *El Moudjahid Economique,* February 3, 1976.

Brugnot, J., "Working Realistically with the Third World Countries in the Field of Education," *Le Figaro,* February 15, 1975.

______ and Plent, A., *Education in France and Europe,* French Social Editions, Paris, 1968.

______, Jahan, A., and Serieyx, H., "Organizing Training for Technicians of the National Public Works Department," *Etudes et Réalisations,* 108, September-October-November 1966.

Cherel, Brouland, Bourdeau, and Vincent, "An Interesting Experiment: Training Saharian Manpower at S.N.P.A.," *Pétrole Information,* 308, May 1961.

Davous, P., "Data Control, Success Factor for Consulting Firms," *Review of the Center for the Advancement of Business Administration;* March-April 1975.

______, "Executives in the Business Enterprise of the Future," *Revue Française de Gestion,* 5, May-June-July-August, 1976.

______, Filiu, M., and de Boisanger, P., "New Frontiers in Strategical Thinking," *Revue Française de Gestion,* special issue, October 1977.

Estrangin, B., "Transfer of Technology, A Real Craftman's Job," *La revue des Ingénieurs,* 247, March 1977.

Eurequip Series, "Languages of Action," Fayard:

—"Introduction to the Computer," 1969.

—"At the Periphery of the Computer," 1972.

—"The Logics of Decision-making," 2 volumes, 1972/1973.

—"13 Criteria for Testing your Judgement," 1972.

—"Introduction to Financial Management" (by George L., and Brugnot, J.) 3 volumes, 1973.

Filiu, M., "At Work, Why Not a Sense of Well-Being? "*Etudes et Réalisations,* 118, May-June, July 1968.

______,"Creativity: Freudian Techniques Serving Business," *Industries et Techniques,* 158, November 1970.

______ and Mathieu, G., "Thoughts on the Meeting between Man and the Computer," *Automatisme,* 6, June 1968.

Garrigues, P., "A Technique to Serve the Future Unitary World," *Synthèses,* 11, 1976.

Giersch, H., "European Industry Faces the Future," 6th European Management Symposium, Davos, 1976.

Gros, A., "A Million Dollars a Day at Stake," *Revue l'Industrie du Pétrole dans le Monde,* 489, December 1977.

Hawthorne, E.P., Technology Transfer, O.C.D.E., Paris, 1971.

Hostache, G., "Quality and Its Management," *Revue Française de Gestion,* 10, May-June/July-August 1977.

Jouven, P., "Communication: the Key to Problems in the Company," *L'Economie,* 1217, June 1975.

Laure, A., "From Urban Planning to the Pluridisciplinary Team," *Prospective,* 11, P.U.F., 1964.

Maguerez, C., "The Technical Advancement of the Illiterate Worker," Eyrolles Publications, Organizational Publications (Business Languages Series), Paris, 1966.

Mariotte, J. Cl., "The Role of Ergonomy in the Creation of a Factory," *Revue des Arts et Métiers,* July-August 1971.

Mathieu, G., "Training Men for Data Processing," *Etudes et Réalisations,* (special issue) May-June 1972.

______, "The Foreman and the Computer," *Travail et Maîtrise,* March-April-May 1970.

Mayer, J. (Satec), and Consigny, V. (Eurequip), "Organizing the Development of a Million People in Madagascar," *Etudes et Réalisations,* 127, 1968.

Miret, P., "Groups and Power in the Business Enterprise," *Revue Française de Gestion,* 4, March-April 1976.

______, "The Case for Training Business Administrators in Social Awareness," *Revue Française de Gestion,* 6, September/October 1976.

Nicolet, J. L., "Control Your Administrative Inflation," *Travail et Méthodes,"* 326/327, June/July 1976.

Okita, S., "Causes and Problems of Rapid Growth in Postwar Japan and Their Implications for Newly Developing Economies," Japan Economic Research Center, Japan, March 1967.

Plent, A., "Screening and Training of Personnel for European Community Nuclear Power Stations," *Bulletin d'Informations A.T.E.N.,* 67, September-October 1967.

______, "The Use of Simulators in the Training of Machine Operators in the Automated Plant," *Revue Automatisme,* 12, December 1967.

______ and Roa, Y., "Teaching Units for Zigsaw-Puzzle Schools," *Etudes et Réalisations,* 106, May-June 1966.

"Removing Barriers to Progress by Adapting Man to Development," *Entreprise,* 661, May 11, 1968.

Reumaux, B., "A Balance Sheet for a Strategical Approach to the Social Problem," *Editions Hommes et Techniques,* April, 1976.

Revol, P. and Vilbert, J. M., "Work-shifts: How to Improve the Working Conditions of Teams Working in Shifts," *Industrie du Pétrole en Europe, Gaz-Chimie,* supplementary issue, 1973.

Sautter, C., *The Price of Power,* Japan, Le Seuil, Paris, 1973.

Serieyx, H., "From Personnel Management to Social Strategy," *Dirigeant,* 59, March, 1975.

______, Reumaux, B., Meunier, A., Constantin, P., and Poujade, M., "Business Reform and Social Strategy," *Dirigeant,* March 1975.

"Setting up SKF in Fontenay le Comte," *Usines d'Aujourd'hui,* 120, May 1973.

Seurat, S., "The Promotion of the Instructor in the Business Enterprise," *C.N.O.F.,* 10, October, 1967.

______, "The New Frontiers of Engineering," *Etudes et Réalisations,* 133, January 1972.

______, "The Ecological Dimension," *Le Management Direction,* 42, December 1973.

______, "Industries in Search of Workers," *Le Monde,* January 28, 1975.

______, "Exporting Knowledge," *Review,* 9, February 1975.

______, "Engineer or Manager?" *Revue Française de Gestion,* 3, January-February 1976.

______, "Turnkey Factories or Businesses," *Investir,* 170, April 1977.

______, "Towards the Exporting of Industrial Capability, *Promotions,* 101, June 1977.

______, "The Exportation of 'Industrial Capability' Also Creates Jobs in France," *Le Monde,* July 26, 1977.

______, "Will We Be Capable of Training Engineers for the Future?" *Les Quatre Vérités,* 43, December 1977.

______, and Plent, A., "From Traditional Education to Career Training," *Etudes et Réalisations,* 106, May-June 1966.

Syntec, "Some Opinions in the Role and Development of Engineering," *Etudes et Réalisations,* 112, 1967.

"Test: Is Your Company Anthropophagous?" *Revue L'Expansion,* April 1969.

Vieux, N. and Bouquand, P., "Ergonomy and Reality," *Journal of the French Association of Petroleum Technicians,* 217, 1973.

Vlad, Z., "Foreign Workers: How to Overcome the Language Barrier," January 1972.

Index